Planes, Jets, & Helicopters

Planes, Jets, & Helicopters
Great Paper Airplanes

John Bringhurst

TAB Books
Division of McGraw-Hill, Inc.
Blue Ridge Summit, PA 17294-0850

FIRST EDITION
FIRST PRINTING

TAB Books is a division of McGraw-Hill, Inc.

Library of Congress Cataloging-in-Publication Data
Bringhurst, John R.
Planes, jets, & helicopters: great paper airplanes / by John R. Bringhurst.
p. cm.
Includes index.
ISBN 0-8306-4451-2
1. Paper airplanes. I. Title. II. Title: Planes, jets, & helicopters.
TL778.B75 1993
745.592—dc20 93-2872
CIP

Acquisitions editor: Jeff Worsinger
Editorial team: Steve Bolt, Executive Editor
Norval G. Kennedy, Editor
Production team: Katherine G. Brown, Director
Jana L. Fisher, Coding
Jana L. Fisher, Layout
Wendy L. Small, Layout
Kelly S. Christman, Proofreading
Stephanie Myers, Illustrator
Stacey Spurlock, Indexer
N. Nadine McFarland, Quality Control
Design team: Jaclyn J. Boone, Designer
Brian Allison, Associate Designer
Cover design: Denny Bond, East Petersburg, Pa.
Cover photograph: Brent Blair, Harrisburg, Pa.
HT1
4426

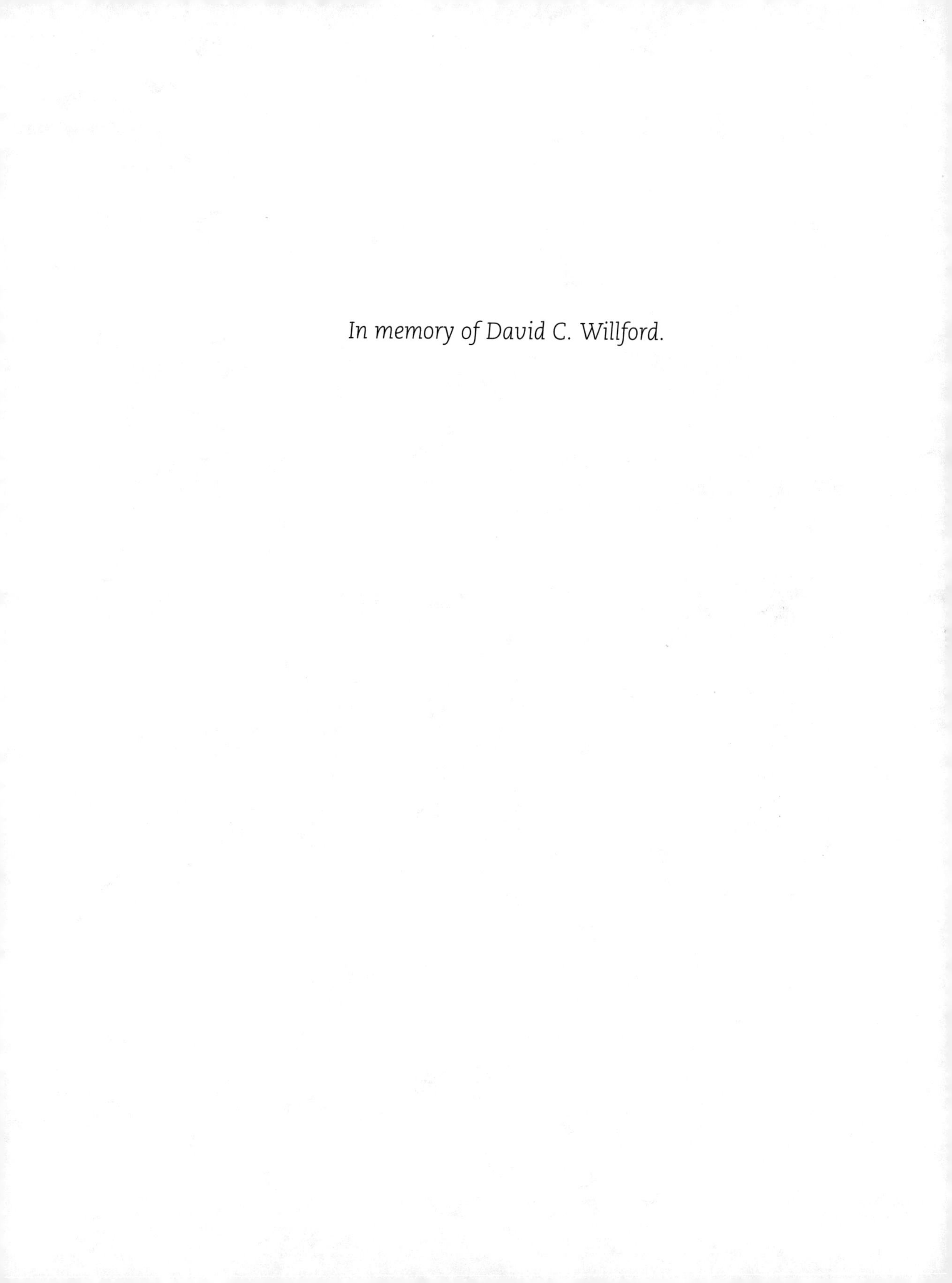

In memory of David C. Willford.

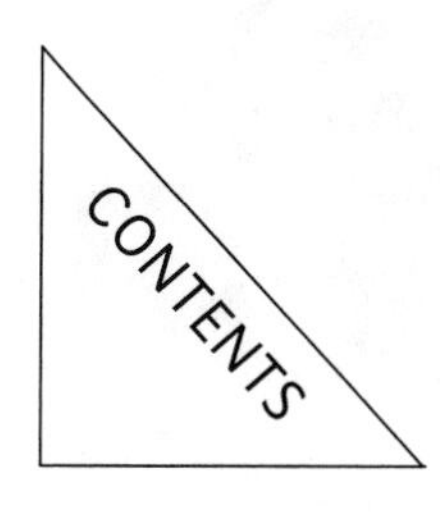

Introduction ix

Planes that look like planes

1 Essential airplane 2
2 Straight wing 6
3 Underside flaps 11
4 Round wing 16
5 Long-winged glider 21
6 Finned wings 25
7 Aerobat 31
8 Midget 36
9 Fuselage plane 41
10 Biplane 50

Jets

11 Hornet 58
12 Aerobatic jet 63
13 Twin fin 68
14 Sleek jet #1 74
15 Sleek jet #2 81

16 Navy jet 87
17 Canard jet 92
18 Stealth flying wing 98
19 Batwing 105
20 Dart 111

Helicopters

21 Basic helicopter 122
22 Broad-winged helicopter 125
23 Rabbit ears 131
24 Missile 136
Index 142

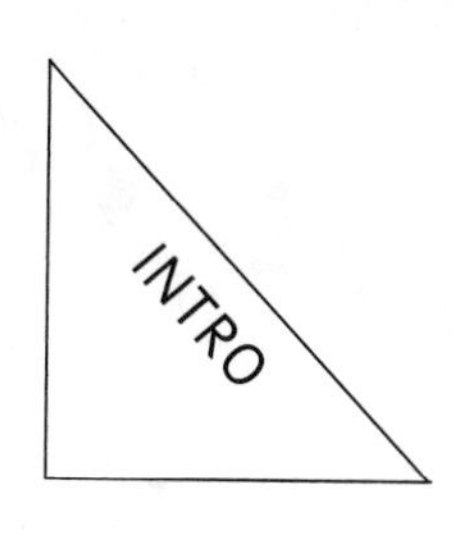

Confessions of a paper folder

CALL ME "DOC."

I have been a closet paper airplane folder for years, plying my hobby secretly by night and when no one was looking, folding shamelessly, folding, throwing, and folding again. Like most kids, I did a little experimenting in grammar school, but when I got to college the urge returned, stronger than ever before. Even as I worked toward an advanced degree, the habit deepened—colleagues never knew of the folding binges I indulged in after a tough exam to "unwind;" nor could they imagine as we sat at lecture that in the dark corners of my mind I pictured a folded masterpiece soaring gracefully, perfectly, over the professor's head.

Yet the secret was there, and after I graduated and entered professional life, it remained like a weighty albatross about my lean neck, unsuspected but ever-present. Even now I confess that after a day at work it is not unusual for me to fold one or two to wind down before bed—sometimes many more.

My wife must have suspected it early on. All the signs were there—the strange appearance of paper planes underneath furniture and behind books in the bookshelf; the unexplained disappearance of every piece of scratch paper in the house; the late nights she endured in bed alone, the silence broken only by the occasional whack of paper against the living room walls. There was the curious discovery of the boxes of planes of every size and description tucked away in our bedroom closet. And then one day it simply hit her in the face (a paper plane, that is, folded perfectly . . . hit her right between the eyes) and she knew. It was a relief, in a way, and she took it in stride. "My husband could have worse hobbies," she reasoned, "like mountain climbing, or skydiving, or gambling. At least this is cheap, and he can do it with the kids." And we have had a kind of understanding ever since—she humors me when I indulge in private, and when the topic comes up in public, she pretends not to know me.

It is now, after years of secrecy and in the spirit of bringing it all out into the open, that I submit the fruits of my self-indulgence: a sampling of the paper planes that have fired my folding habit these many years.

Of a truth, the addiction has been a modest one; here you will find, I think, no breakthroughs in aerodynamics, yet neither will you find the demand for superhuman patience and precision; only the bare honesty of paper, 8½-x-1-inch paper, the paper that is the hallmark of our society, the paper you see overflowing from recycling bins, wastebaskets, and copy machines, paper waiting, begging to be put to a higher purpose. The appeal of paper airplane folding is that it is a kind of aeronautical engineering for the common man—it is easy, it is fast, it is inexpensive, it doesn't hurt anyone when things go awry, and the rewards of a successful flight are indescribable; sort of a momentary compensation for those unfulfilled childhood dreams of becoming a test pilot. Try it—a little bit can't hurt.

The rules

All the airplanes in this book follow three basic rules. I like to think I invented these myself, but in truth I think they have been there as long as paper airplanes have been folded, and I merely follow them, like anyone else. Certainly to depart from them would be to open a Pandora's box of possibilities, thereby bringing chaos to an otherwise tidy art.

Rule #1

Each plane must be folded from a single sheet of paper: 8½ × 11 inches. Exhaustive research has demonstrated this to be the ideal size and shape for optimum aerodynamic performance; besides, these happen to be the dimensions of standard letter-sized paper, which is available everywhere, and most importantly, is very cheap, and can often be had for free. That convinces me.

Rule #2

Planes must be folded only: no cutting, taping, or gluing. I follow this rule as a matter of practicality—I am a little dangerous with a pair of scissors, and glue takes too long to dry and makes fingers stick to the paper. Besides, I have always regarded as cheating the use of adhesives of any kind. If we're taking the trouble to fold the plane rather than buying a kit (or better stated, avoiding the trouble of the latter by doing the former), the folding should hold the plane together, should it not? Enough said.

Rule #3

Every plane must fly, or at least do something interesting in the air. This should be obvious to everyone, though some who advocate art for art's sake might find it boorishly confining. Nevertheless, rules are rules; if it doesn't fly, it doesn't belong.

When you think about it, these rules are nothing more than those that prevailed on the playground where most of us folded our first paper airplanes—we used the paper that was available, folded it in the simplest way possible, and of course we made it fly. That is, after all, the object of the undertaking.

Paper folding basics

It is my hope that most readers will skip over this preliminary material, open the book to a plane they like, and simply start folding—most will find the instructions and illustrations to be self-explanatory, as long as the directions are read and followed carefully. This section is intended for those who have difficulty transforming the instructions into actual folds on a sheet of paper. Recognizing that some people have been tragically deprived of the joys of paper airplane folding and hence might be unaware of the basics, we will first discuss some general guidelines of paper folding simply to make life easier, then follow with some specific types of folds that are frequently used.

General guidelines

- Fold your plane on a smooth, flat surface, such as a table or large book. It is very difficult to try to fold on an uneven surface, and nearly impossible using only your hands without a flat surface to fold against.

- Align each fold carefully before creasing, then crease with a smooth motion using your fingertips against the folding surface. If the fold takes the wrong path, you can usually realign it and crease again.

- Follow the directions carefully. Some illustrations have several instructions, and each instruction must be completed before moving on.

- On most planes, the exact placing of any fold is less important than that the folds be made equally on both sides of the plane.

- Be as precise as you can, but don't worry if some folds aren't perfectly accurate—in most cases the plane will fly anyway.

- It is best if you use good paper (these designs were tested with photocopy-quality bond paper), but almost any standard-sized paper will work for most designs, even three-hole binder paper. Generally, the paper you use should be fairly smooth, without folds or crumpled areas.

Folding against an edge

Many of the airplane folds are made using a crease or an edge as a guide. Figures 1–5, for example, show a series of two consecutive diagonal folds against a vertical crease. Note that each fold extends from the top edge where the vertical crease ends; this is where you should start your fold. First, begin the fold at the point marked A in Fig. 1, and holding that position down with your finger, align the top edge AB (between A and B) with the vertical crease, as shown in Fig. 2. Keeping this edge aligned, flatten down and crease the fold from point A diagonally outward to point C, as shown by the arrow.

The second fold is made the same way. The fold is started at point A,

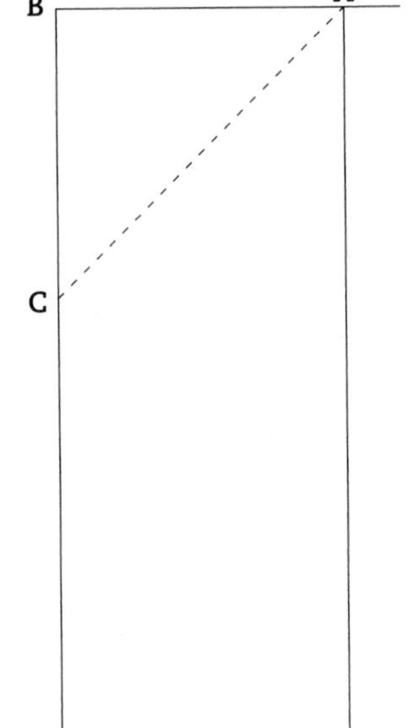

Fig. **1** *Folding against a crease.*

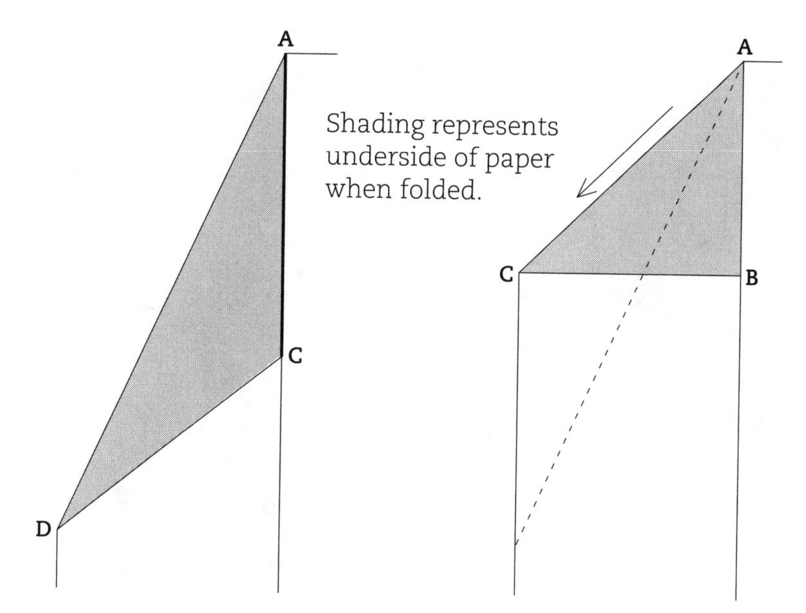

Fig. **2** *Folding against a crease.*

Fig. **3** *Folding against a crease.*

and edge AC is aligned against the central crease as shown in Fig. 3, then holding that edge firmly aligned with one hand, the folded section is flattened down and creased firmly from point A out to point D. This type of fold might take some practice at first if you have not done much paper folding, but it quickly becomes a simple matter.

Sometimes, especially in heavily folded designs, the thickness of the folded paper edges can make further folding difficult; therefore, when making the folds against a crease as shown in Figs. 1–3, it sometimes helps to align the edge a very short distance away from the crease and parallel to it, rather than exactly on it. If you then have to fold along the crease as shown in Figs. 4 and 5, there will be sufficient room for the bulky folded edges and the fold will be much easier and more tidy.

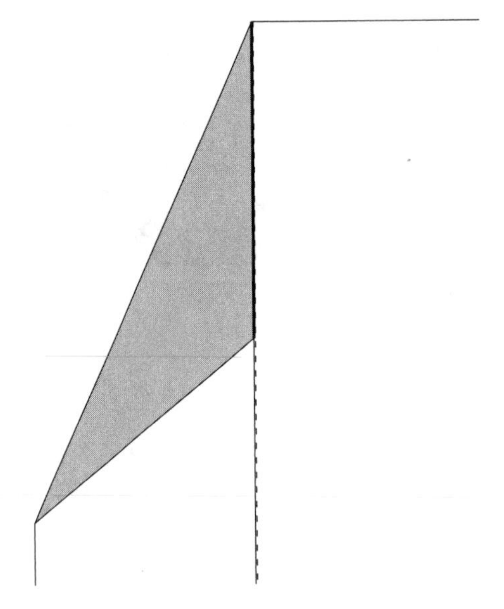

Fig. **4** *Folding against a crease.*

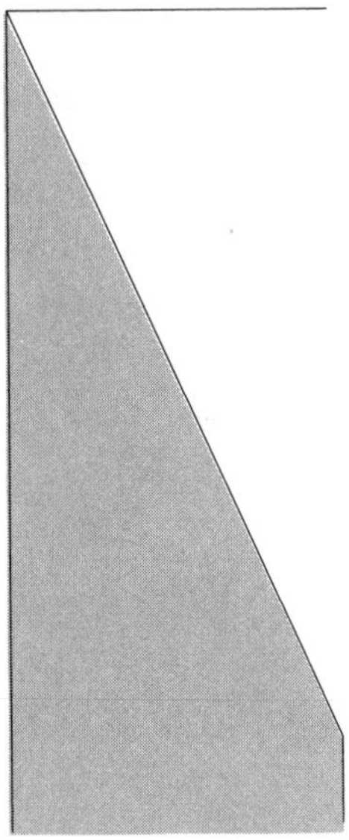

Fig. **5** *Folding against a crease.*

Folding from point to point

Some folds are created by aligning one point with another point and folding crosswise between them. This type of fold is illustrated in Figs. 6–7. This fold is formed by aligning the tip, labeled A, with point B where the horizontal crease crosses the vertical one. Note that this creates a fold exactly between these two points. Holding the tip down against point B as shown in Fig. 7 with one hand, you first crease flat at point C in the center, then extend the crease to both sides as shown by the arrows.

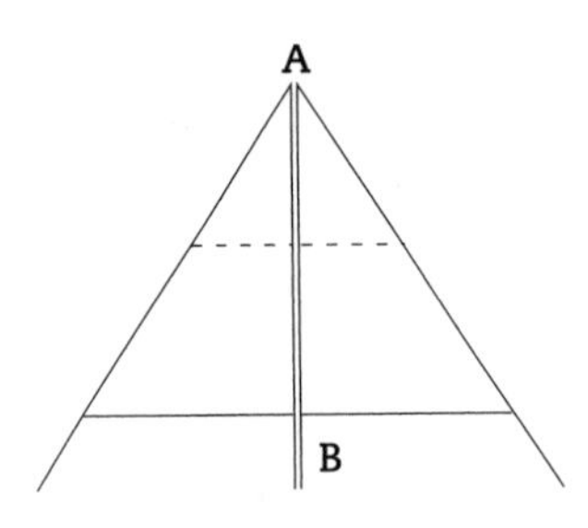

Fig. **6** *Folding from point to point.*

In making such a fold, it is important to keep the center of the plane exactly aligned as shown in Fig. 8. If the fold is made unevenly, as in Fig. 9, subsequent folds will also be uneven and the plane will likely fly poorly.

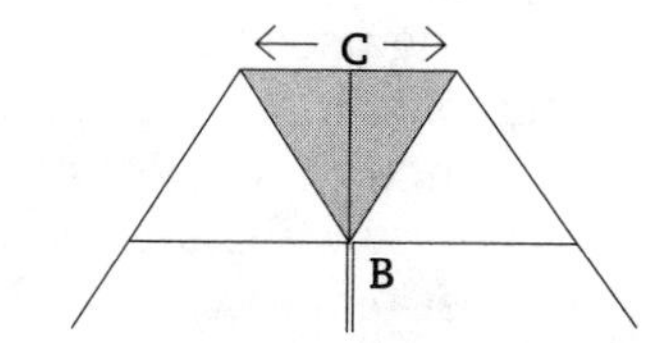

Fig. **7** *Folding from point to point.*

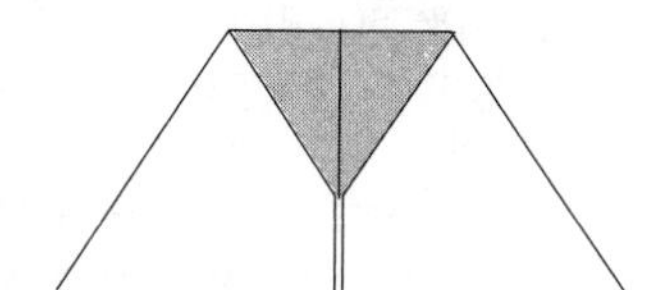
Fig. **8** *Folded correctly.*

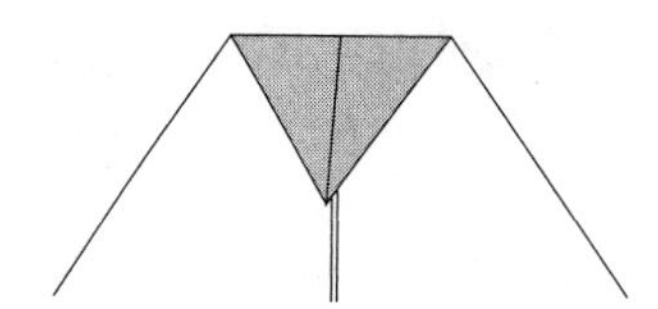
Fig. **9** *Folded incorrectly.*

Accordion folds

Certain planes require one or more accordion folds—that is, folds on which a segment is turned inside-out like the pleats of an accordion. Accordion folds are actually quite simple to form, and an example is shown in Figs 10–13. Figure 10 shows part of the tail end of an airplane from the side, with the wings folded upward. To form the accordion fold, the end of the tail, marked A, is pushed upward between the wings, causing the folded section AB to turn inside-out, as shown by the dotted section in Fig. 11. The folds from B to C should then be flattened from the sides and creased. Figures 12 and 13 show the same fold from above, with the wings extended.

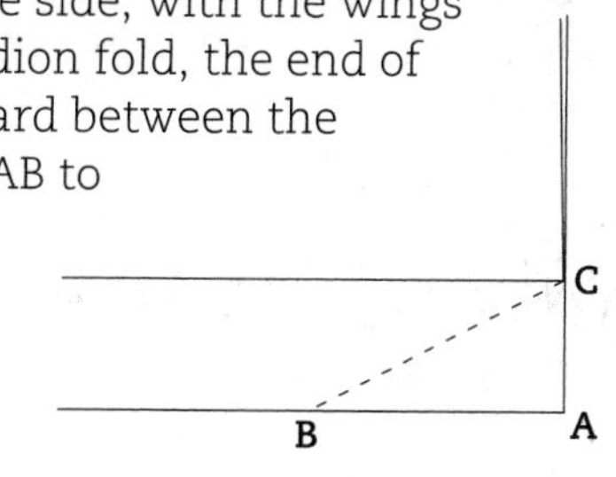

Fig. **10** *Accordion fold.*

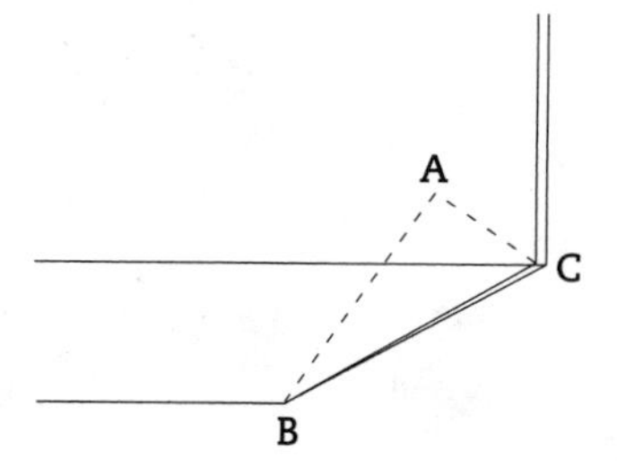

Fig. **11** *Accordion fold.*

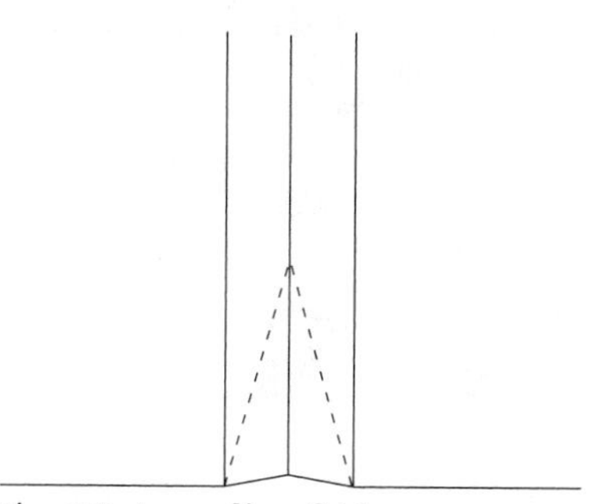
Fig. **12** *Accordion fold.*

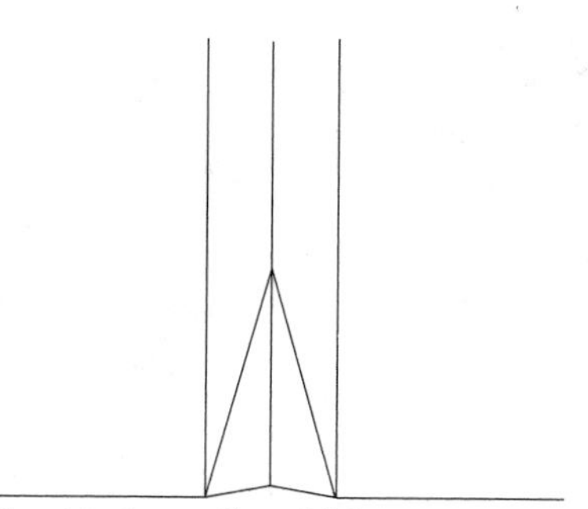
Fig. **13** *Accordion fold.*

A few planes require more complex accordion folds, or even complicated telescoping folds; instructions are given for these folds where needed.

Making it fly

Paper has a personality that takes some getting used to, and occasionally a paper plane, however nicely folded, refuses to fly well. Instructions for many of the airplanes include suggestions for adjustments in case the airplane does not fly well; however, the following general guidelines apply to most planes.

Proper throw

The way a plane is thrown often determines the way it flies. A plane should be thrown with a straight pushing motion (Fig. 14), as you would throw a dart, rather than a curved motion as you would use to throw a baseball (Fig. 15). The speed of the toss is also important; for example, a plane resembling early straight-winged planes will not fly well with a supersonic launching. Each airplane has a "To Fly" paragraph that provides specific instructions on how best to fly that particular plane. A good flight might be a matter of a little practice.

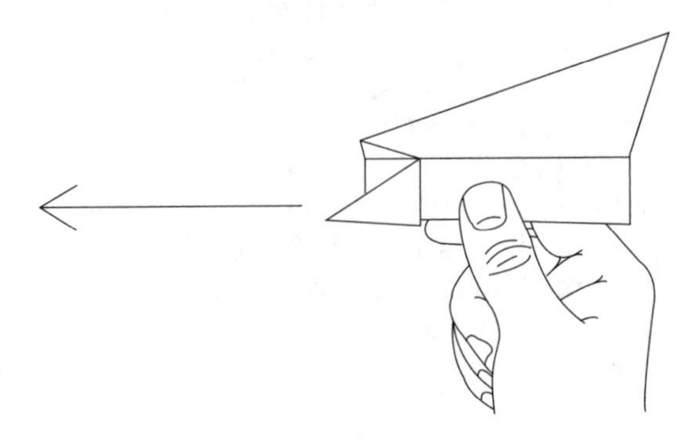

Fig. **14** *Correct toss.*

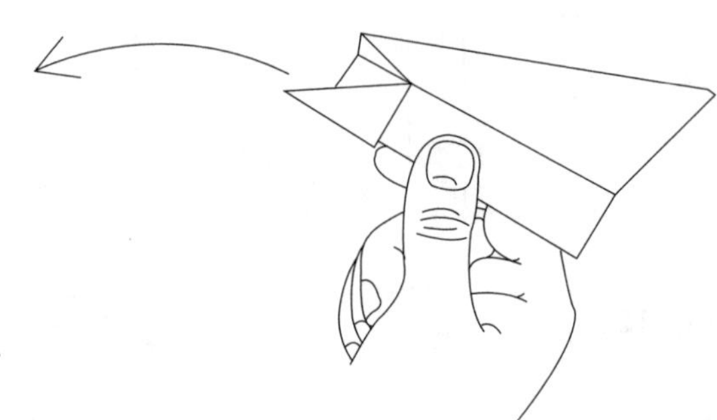

Fig. **15** *Incorrect toss.*

Proper shaping

The plane should be symmetrical—right and left sides should match as exactly as possible. Sight along the center of the plane from back to front to see if the wings match in most details, and make necessary adjustments. It sometimes helps to fold the wings against each other to realign them, and to carefully refold all the edges so they are sharp and exact.

Proper steering

If with careful folding the plane still veers off course, try steering the airplane. Steering a plane is a little like steering a boat, except that whereas a rudder is used behind a boat to steer it right or left, on a paper plane the back edge of the wing is bent to steer the wing up or down. (In some planes, the entire wing can be tilted upward or downward.) This bend is called an *elevator* or *aileron*, and is made as follows:

- If the plane dives into the ground or descends too quickly, bend the back edge of both wings upward slightly. This will cause the plane to climb more.
- If the plane noses up and stalls, do just the opposite—bend the back wing edges down slightly until you get a steady flight.
- If the plane veers off to one side, observe which wing goes up. Slightly bend upward the back edge of that wing, which should make the errant wing tend to dip in flight, straightening the turn. Make another short test flight, readjusting if necessary.

If, with some effort at adjustment, the plane still won't fly, why not just get another piece of paper and try again? It takes little time and effort—that is the beauty of paper airplane folding, after all. Follow the instructions carefully and crease well, and your plane should fly.

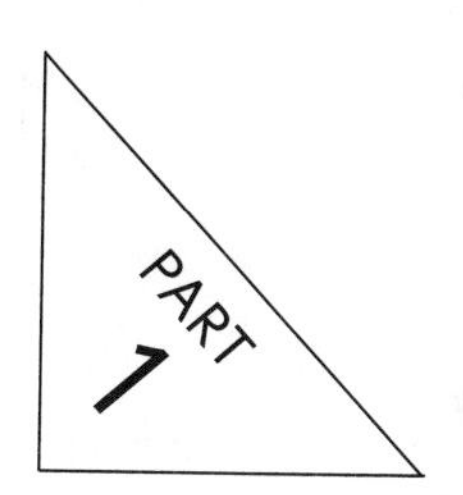

Planes that look like planes

THE FIRST self-propelled airplanes of our day were just what the name implied—essentially flat surfaces that were pushed through the air to achieve flight. While the airplane has undergone tremendous changes through the course of the century, the mention of a plane still evokes the image of those early designs with straight, square wings and slow, predictable flight. Paper plane designs with those flight characteristics comprise the first part of this book.

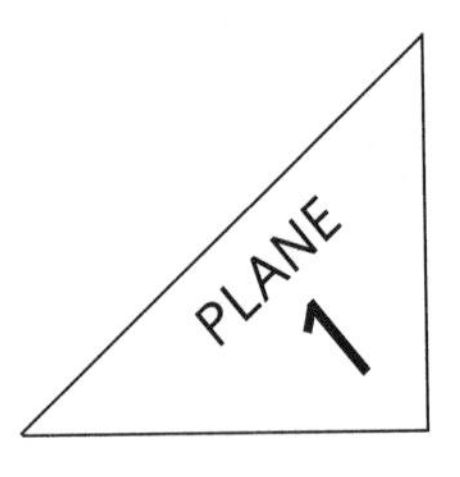

Essential airplane

THIS FIRST DESIGN is nothing more than the airplane stripped down to its most fundamental form—that of a straight flat surface with an *airfoil*, which is a lift-producing curve. Although this plane sometimes requires a little adjustment, it flies amazingly well and demonstrates how little is really required to achieve flight. (Folding instructions are on pages 3–5.)

To fly

Hold the plane gently from behind so that the folded edge is forward and the flaps down, as in Fig. 10. Your middle three fingers should rest on top of the wing, with the thumb and pinkie beneath. Push forward horizontally, and release.

Adjustments

If the plane dives into the ground, try sliding your finger forward along the underside of the wing, between the wing and the front fold, to edge the fold slightly away from the wing. Or lift upward on the back edge of the wing in the center to create an upward bulge. If the plane veers to one side, try adjusting the angle of the side flap on either side. Occasionally this plane will fly best upside down, which is fine so long as it flies.

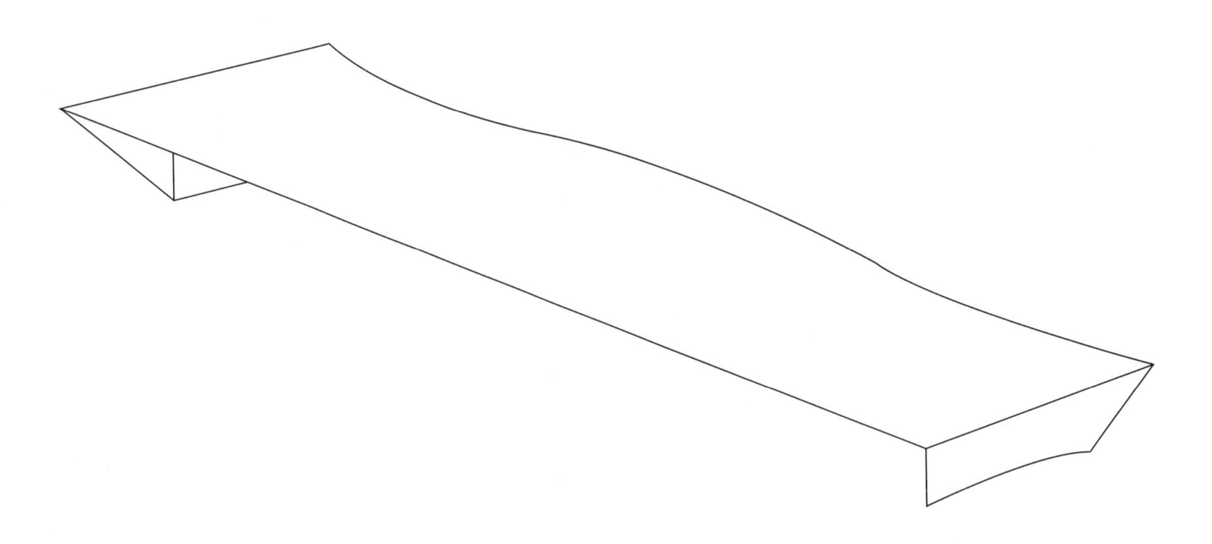

1. With the paper as shown, make a central horizontal crease by aligning the top edge of the paper with the bottom.

Unfold.

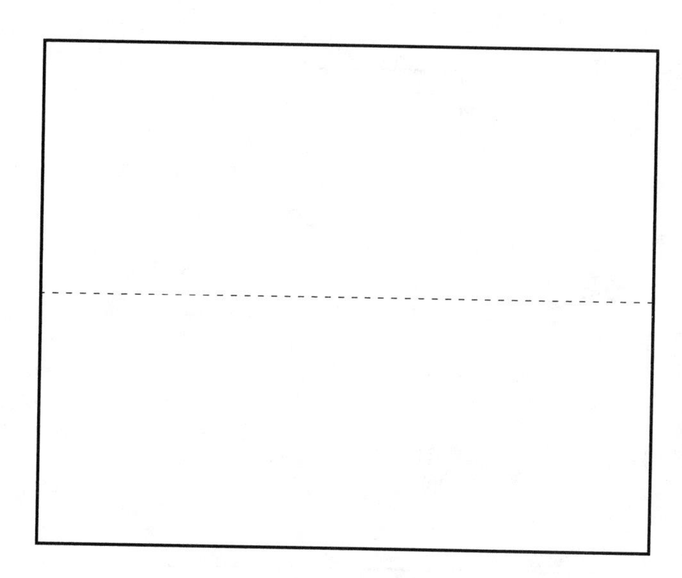

2. Fold down opposite corners as shown, carefully aligning the side edge from each corner with the central crease you just made. *See* Fig. 3.

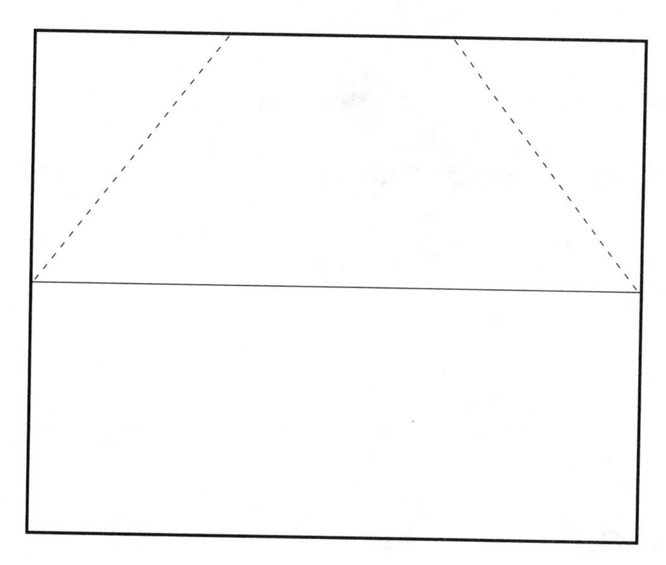

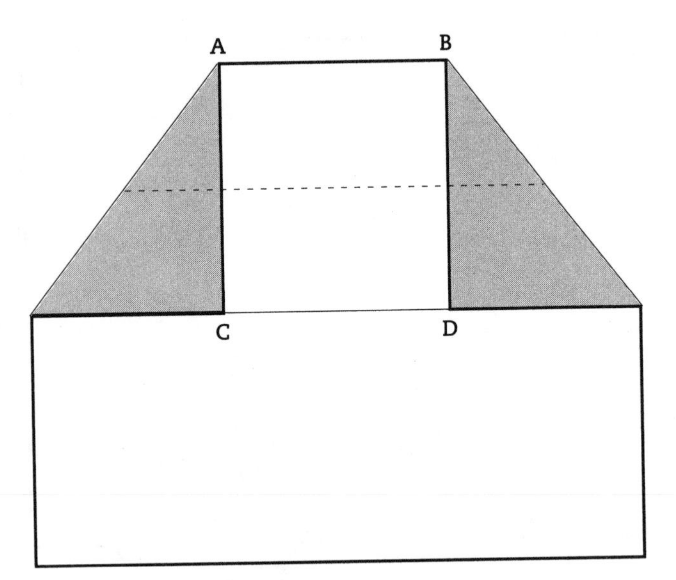

3. Fold the top edge of the paper down, aligning it against the central fold. Points A and B of the top edge should exactly meet points C and D on the central fold.

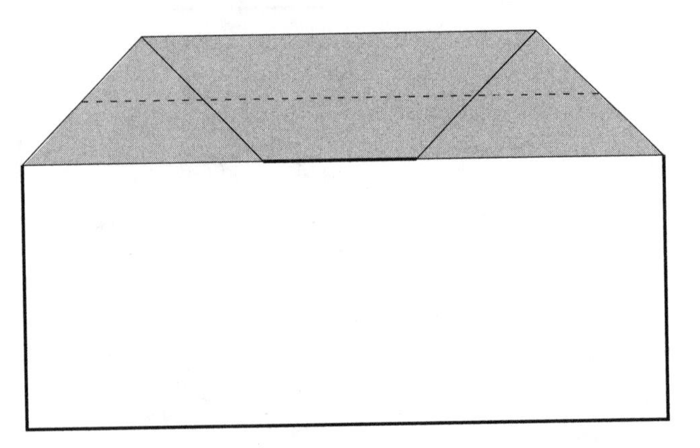

4. Again fold the top edge against the central crease.

5. Fold over along the central crease.

6. Fold down the top corners as shown, matching point E to point F on each side.

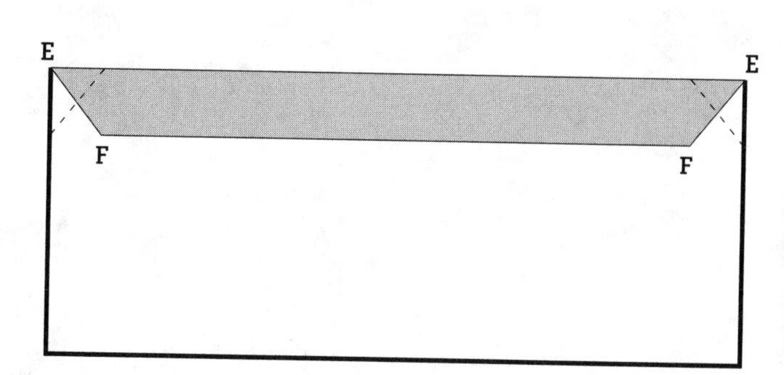

7. Fold down just below the folded edge, as shown.

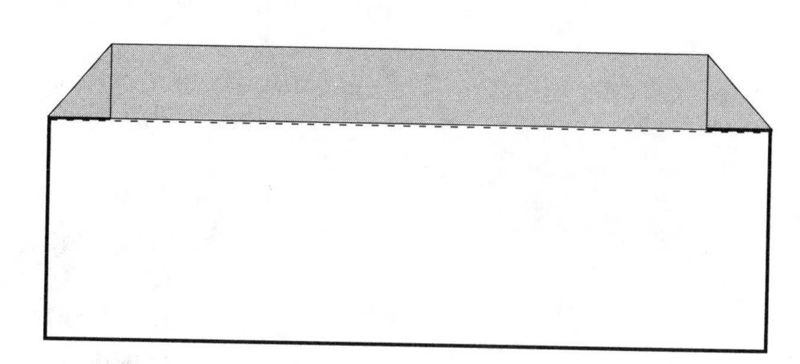

8. Again fold down the top corners, as shown.

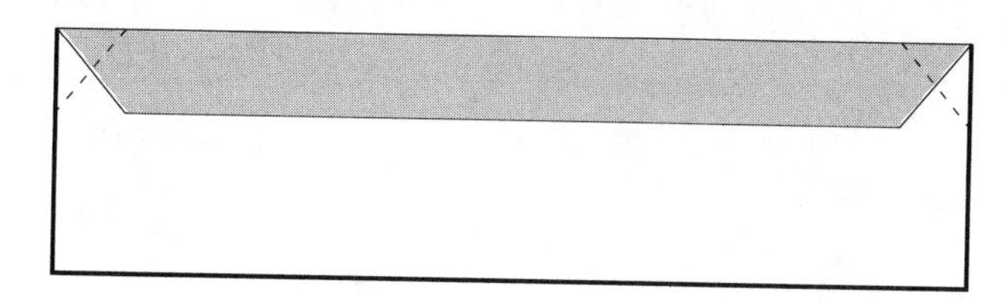

9. Fold the side edges up so that the folds pass through the corners just formed.

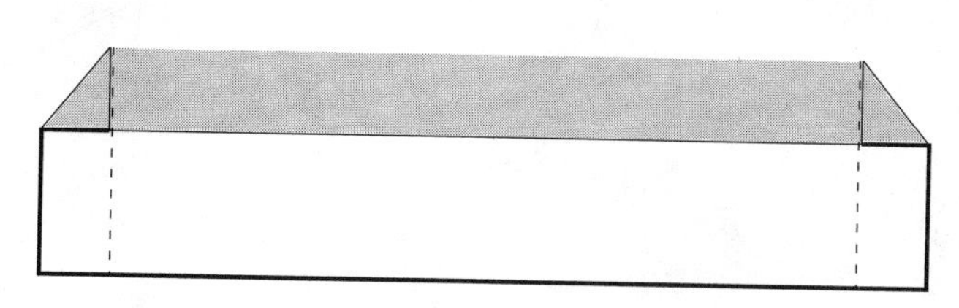

10. The end flaps should form an acute angle with the body, as shown on this front view.

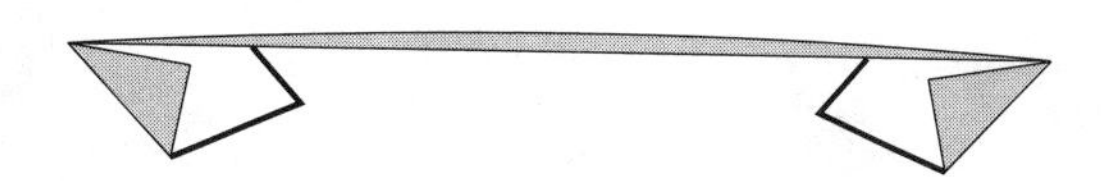

Flying instructions for PLANE 1 are on page 2.

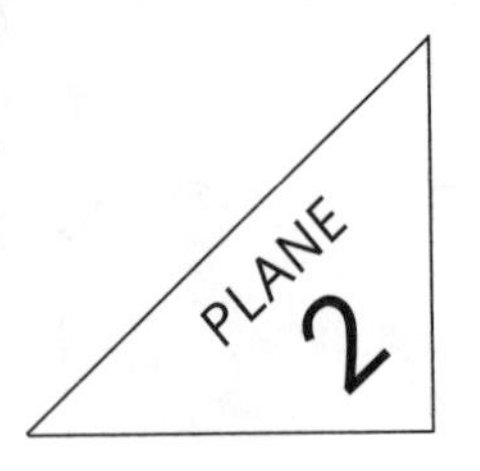

Straight wing

WITH ITS PERFECTLY rectangular wings and rudimentary tail, this plane is reminiscent of some of the early experimental crafts from the days when aviation was in its infancy. Its smooth, slow flight makes the resemblance even more striking. (Folding instructions are on pages 7–10.)

To fly

Hold the plane by the body from beneath, with the tail to the back. Push forward smoothly, and release.

Adjustments

Note that the upper surface of the wings are gently curved. This curve is the airfoil, which helps the plane achieve lift in flight. If necessary, increase this curve by running your finger along the underside just behind the front edge. Make certain that both wings have approximately the same appearance, adjusting as needed. This plane flies best if the wings in flight are very nearly horizontal.

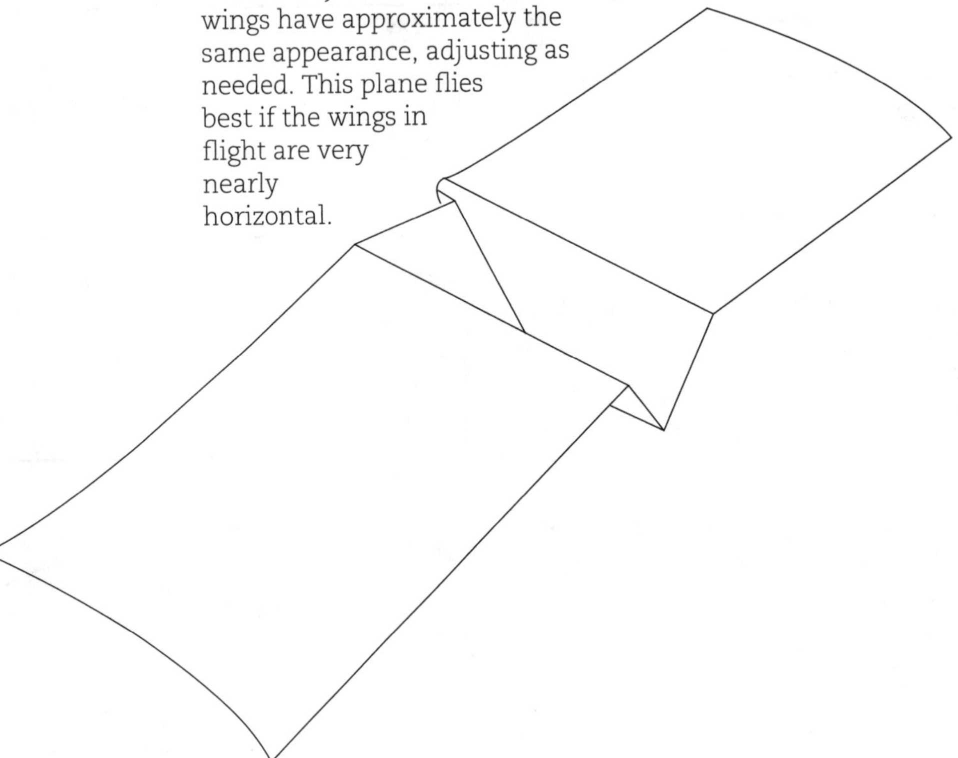

1. Fold the paper in half as shown, aligning the side edges exactly. Crease well.

Unfold.

Turn the paper over.

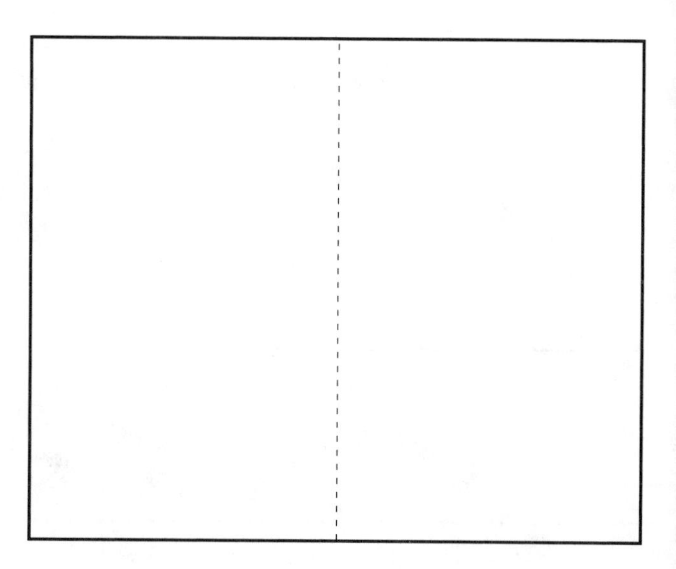

2. Fold down the two upper corners, aligning the top edge exactly with the central crease. The result should look like Fig. 3.

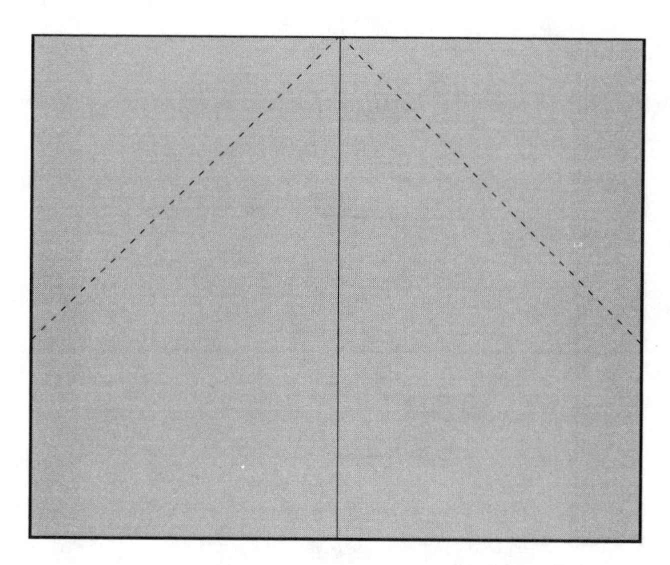

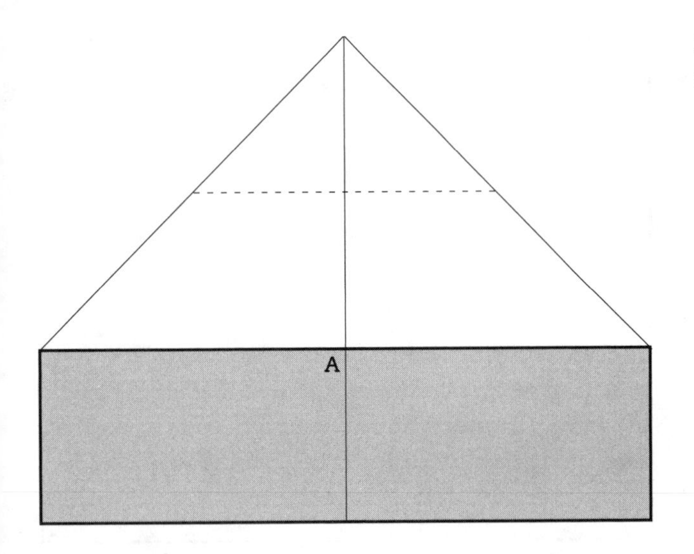

3. Fold the tip down to the point A as shown in the drawing.

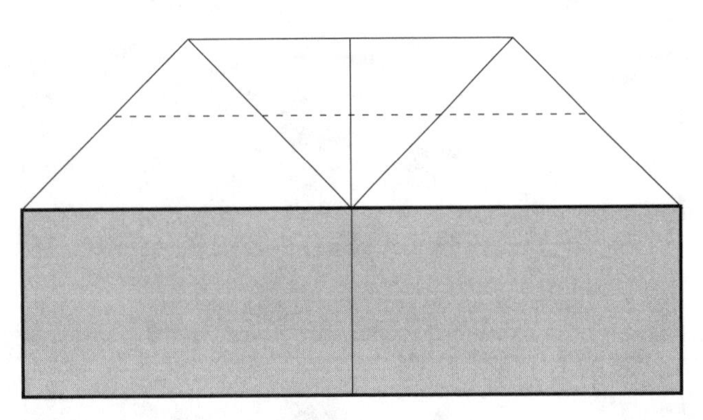

4. Fold the top edge down to the horizontal paper edges, as shown. Be sure to keep the central crease aligned.

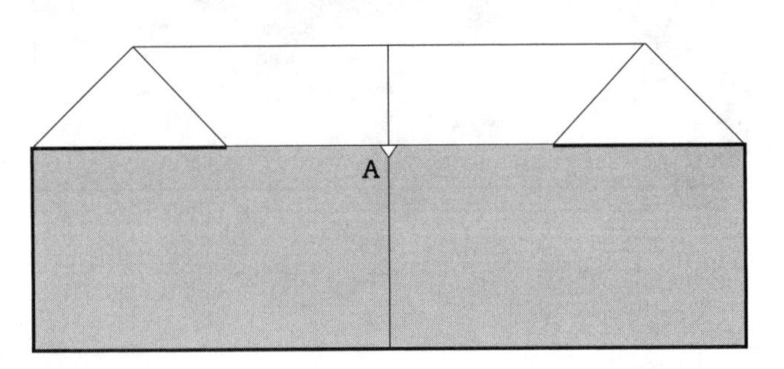

5. Shows the result of the previous fold. Note that at point A a corner of folded paper is found just under the resulting folded edge. Grasp this, and pull it downward, unfolding it out.

6. The result should look like this. Now refold this section of paper by pushing the fold BC all the way forward into the space underneath the paper edge between point D and point E. The result should look like Fig. 7.

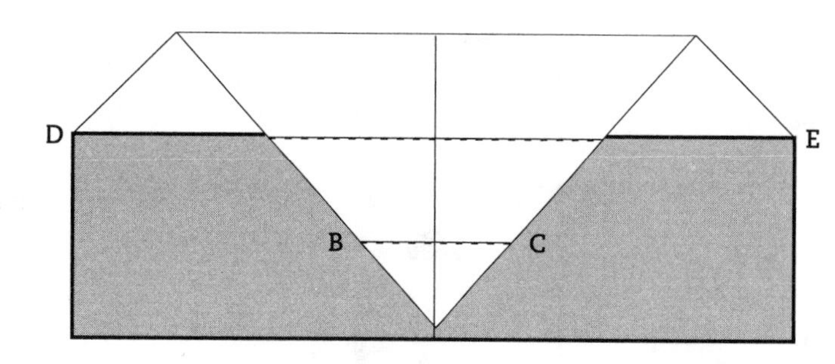

7. Now fold down the upper corners so that the diagonal edges on each side lie against the horizontal paper edge, as shown in Fig. 8.

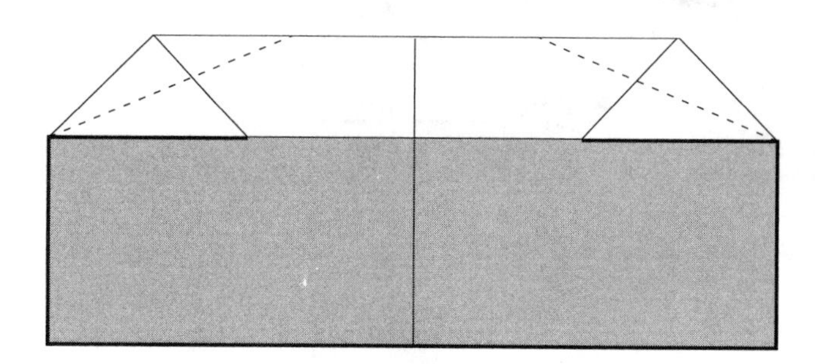

8. Fold down the front of the plane along the horizontal edge shown.

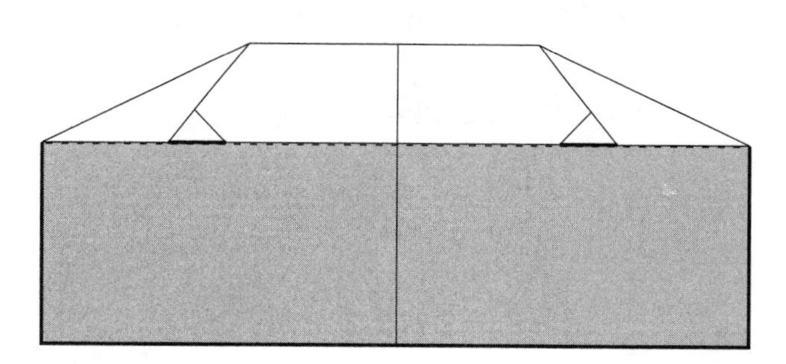

9. Shows the result of Step 8. Turn the plane over.

Fold in half along the central crease, taking care to keep the wings exactly aligned.

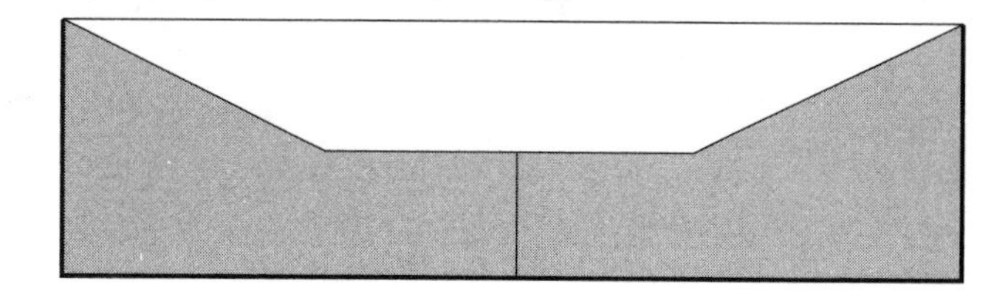

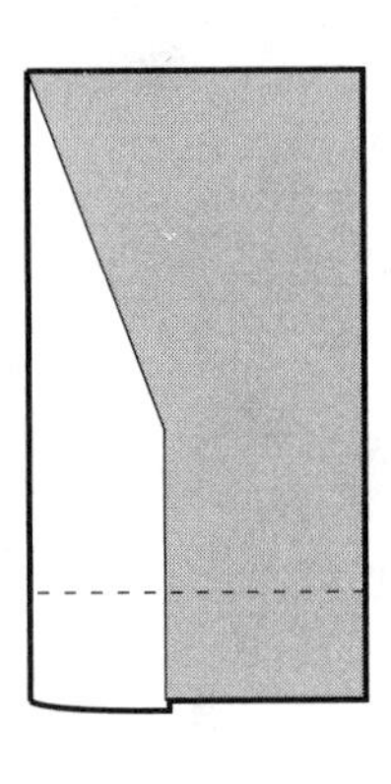

10. Fold down each wing, so that the base of the wing is parallel to the central fold and about 1 inch from it. See that both wings are folded in exactly the same place.

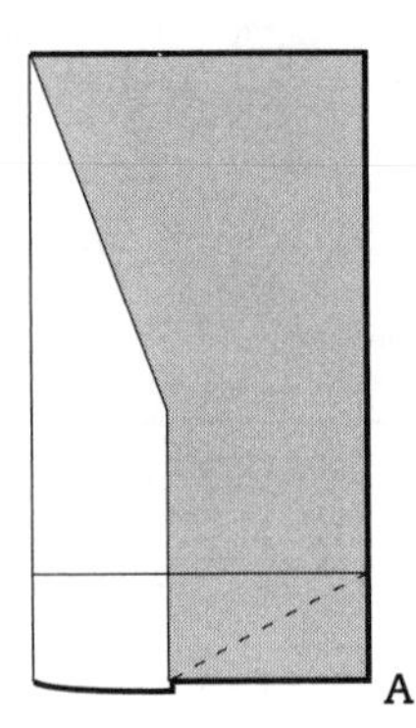

11. This is a side view with the wings bent upward. Form the tail by making an accordion fold as shown, bringing tip A upward between the wings, turning the central fold on the tail inside out, and creasing as shown from the sides.

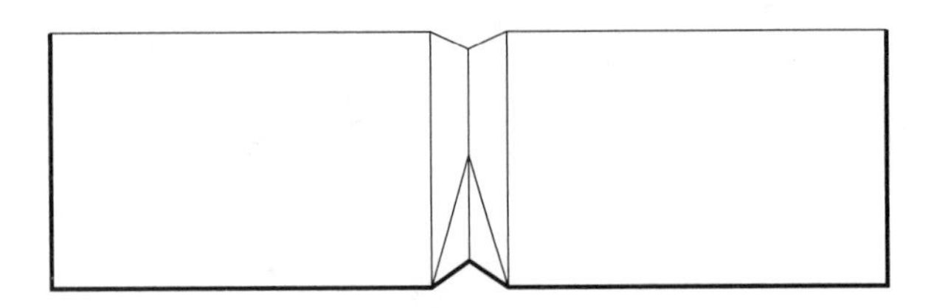

12. This shows the resulting plane, from above.

Flying instructions for PLANE 2 are on page 6.

Underside flaps

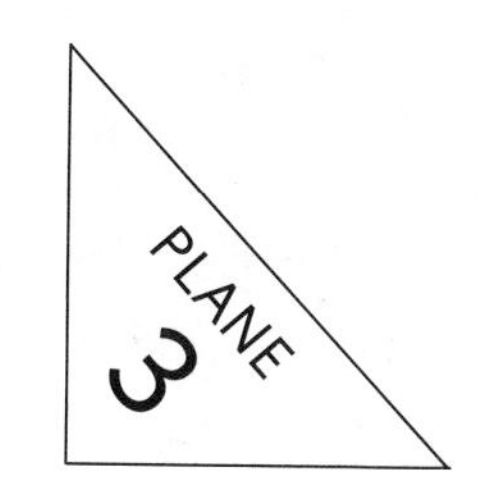

THIS PLANE OWES its smooth, stable flight to a very unusual feature—small flaps folded directly down from the underside of each wing. Without the flaps, the plane dives into the ground (as you can see for yourself by throwing it prior to completing the final step); with the flaps, the plane flies perfectly. (Folding instructions are on pages 12–15.)

To fly

Hold the body from underneath, the folded wing edge forward, and give a firm horizontal toss. It is interesting to try flying the plane before and after step 10, to observe the results of the flaps under the wings.

Adjustments

If the plane tends to turn to one side, you can correct the tendency by slightly flattening down the underwing flap on the same side as the direction of the turn, or by slightly increasing the angle of the flap on the opposite side.

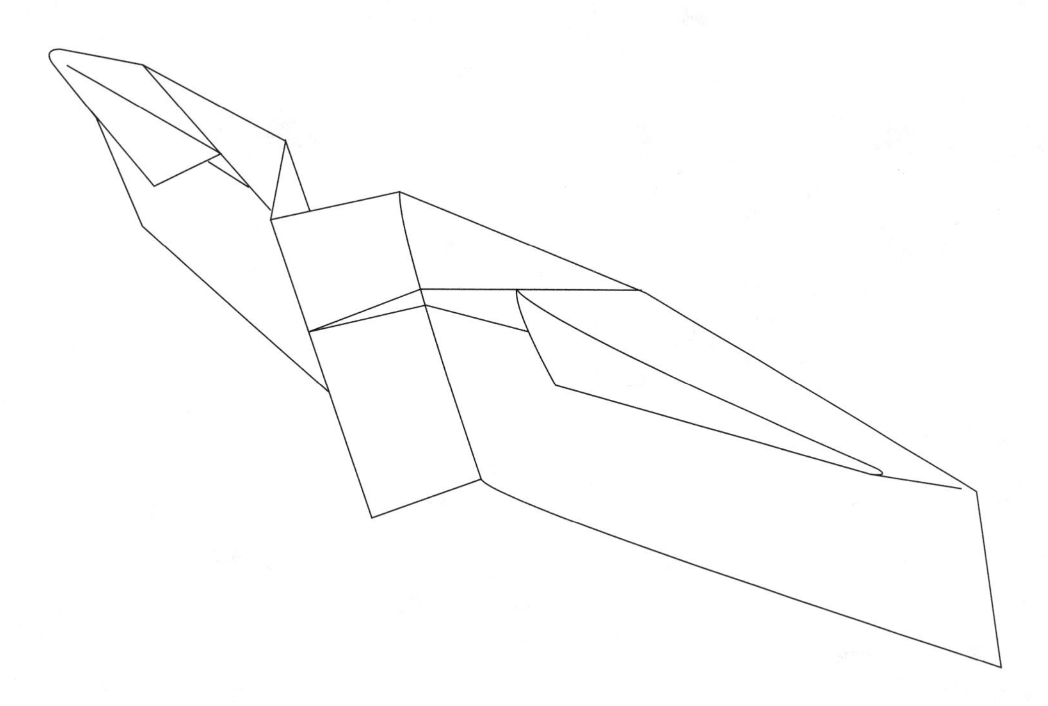

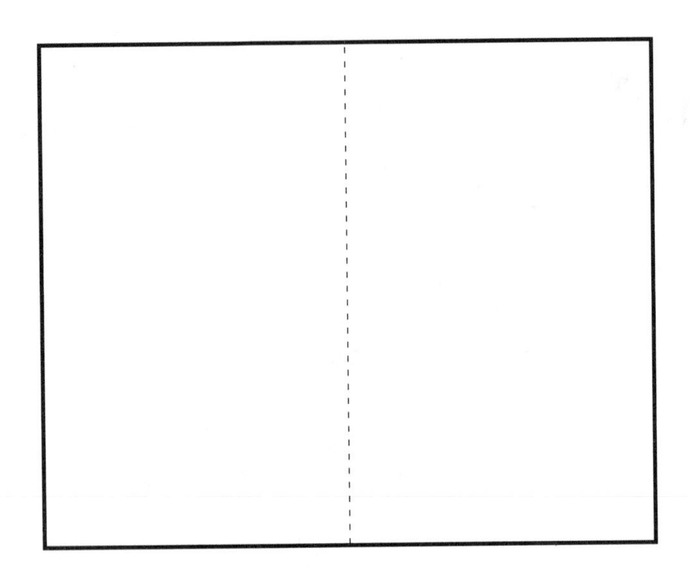

1. With the paper as shown, make a vertical crease by aligning one side edge of the paper with the other.

Unfold.

Turn the paper over.

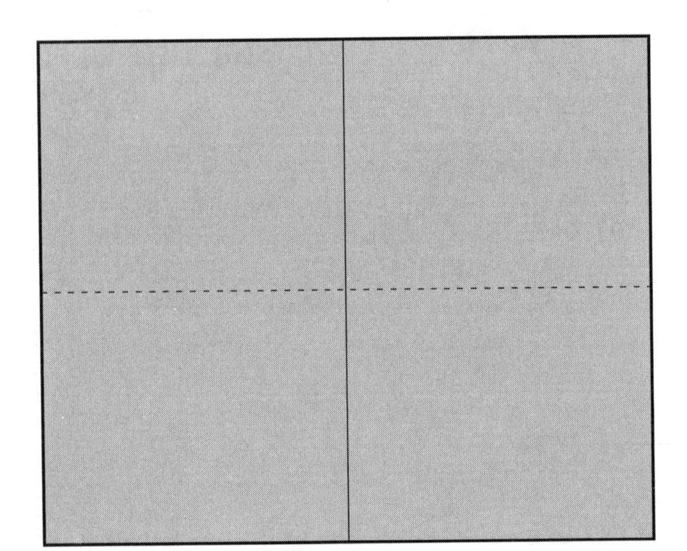

2. Make a horizontal central crease as shown, by aligning the top edge exactly with the bottom edge. Be certain the vertical crease is also aligned before creasing.

Unfold.

3. Fold the top edge down flush with the horizontal crease.

4. Fold again, along the horizontal crease, as shown.

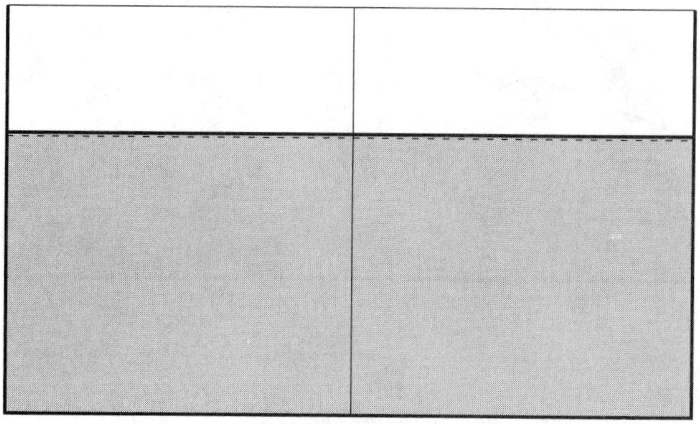

5. Fold down opposite corners, as shown. They should look like Fig. 6.

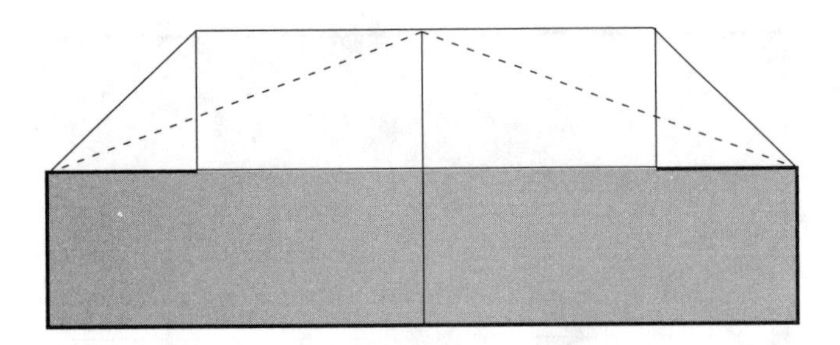

6. Fold down each side so the diagonal edges align with the horizontal edge, as shown. *See* the result in Fig. 7.

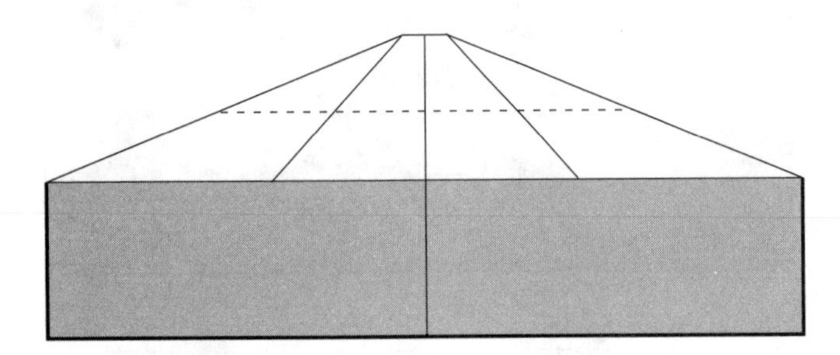

7. Fold down the top to align with the horizontal edge. Keep the central crease aligned.

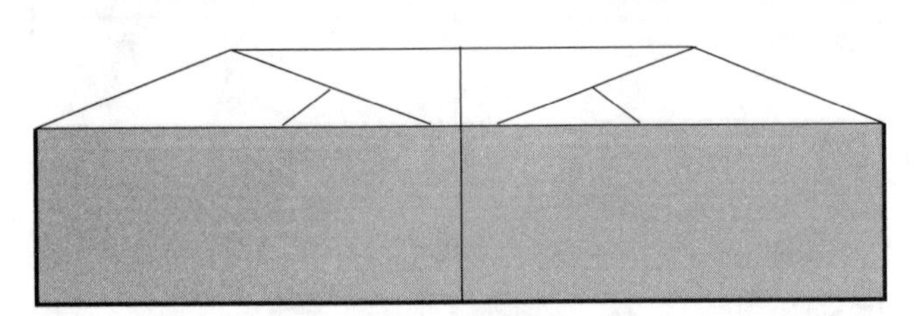

8. Shows the resulting plane from underneath. Turn the plane over.

Fold in half along the central crease. The wings should align exactly with each other.

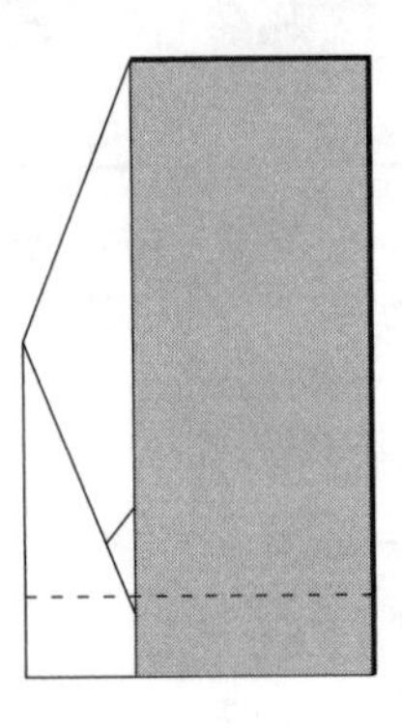

9. Fold down each wing as shown, approximately ¾ inch from the central fold. Be certain to fold both wings exactly the same. Carefully recrease the fronts of the wings, so the crease is fairly sharp.

10. Place the plane bottom-up as shown. Make the underwing flaps by folding up the corners shown on the underside of each wing. Leave the flap sticking out and slightly rearward under the wing.

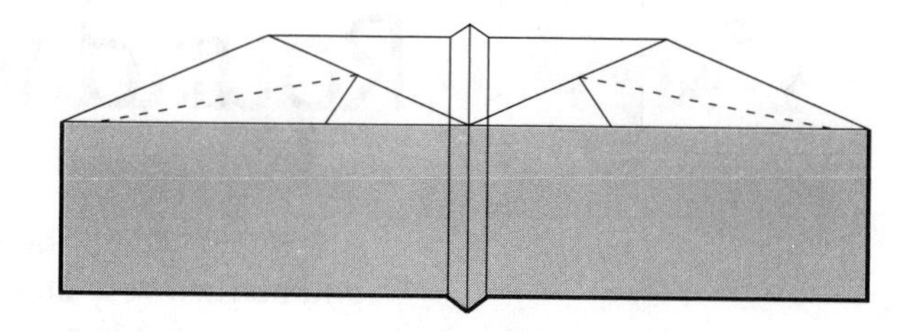

Flying instructions for PLANE 3 are on page 11.

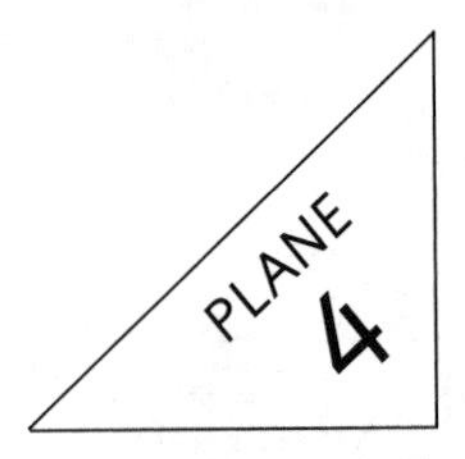

Round wing

THIS PLANE USES the same underwing flap principle as PLANE 3, and so flies very smoothly. The multiple angles on the front of the wing give the impression of roundness and make it pleasing to the eye. (Folding instructions are on pages 17–20.)

To fly

Hold the plane from beneath and give it a strong throw directly forward.

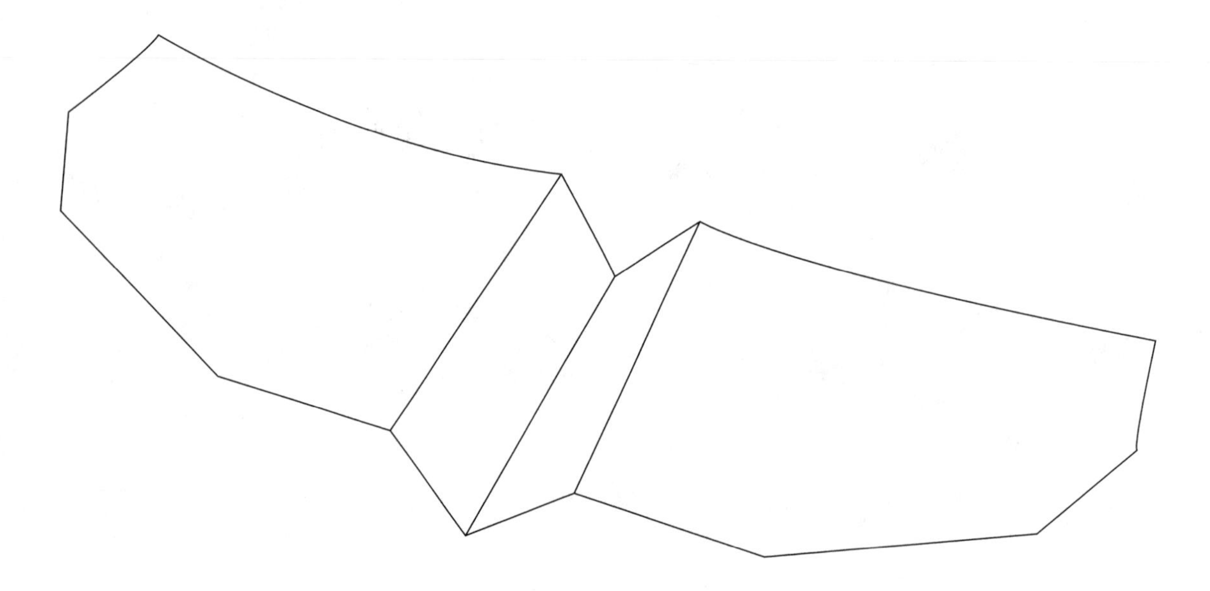

1. With the paper as shown, make a vertical midline crease by aligning the left edge of the paper with the right.

Unfold.

Turn the paper over.

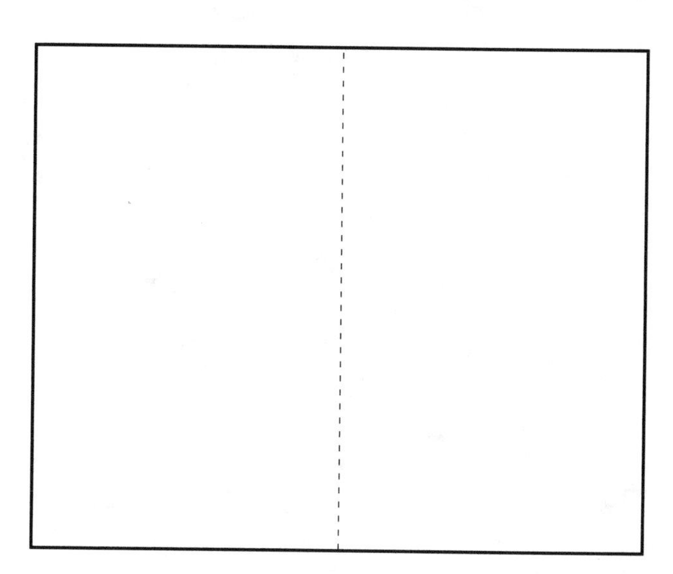

2. Make a small crease about one-third from the top of the paper by bringing the top edge down until it appears about halfway to the bottom edge, and creasing with one finger across the central crease.

Unfold.

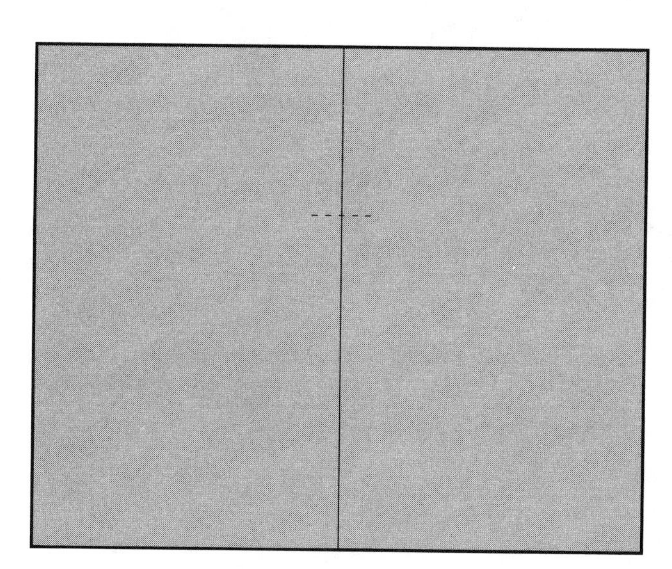

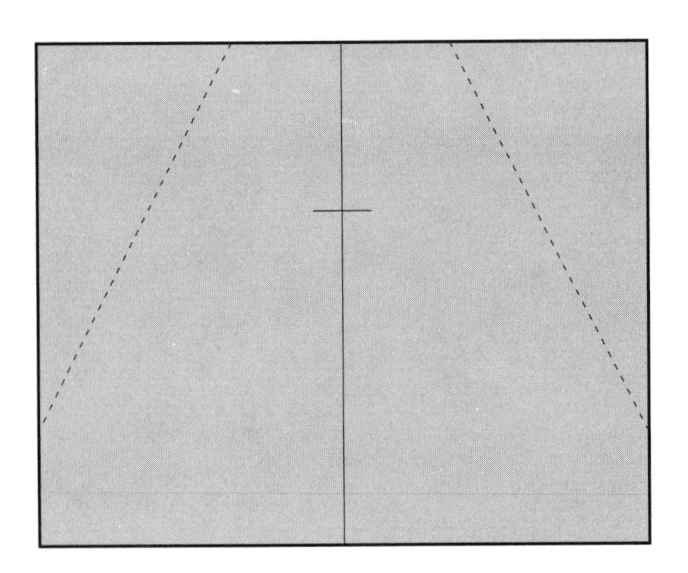

3. Now fold down the right top corner so it just touches the intersection of folds you just created. Repeat with the left top corner.

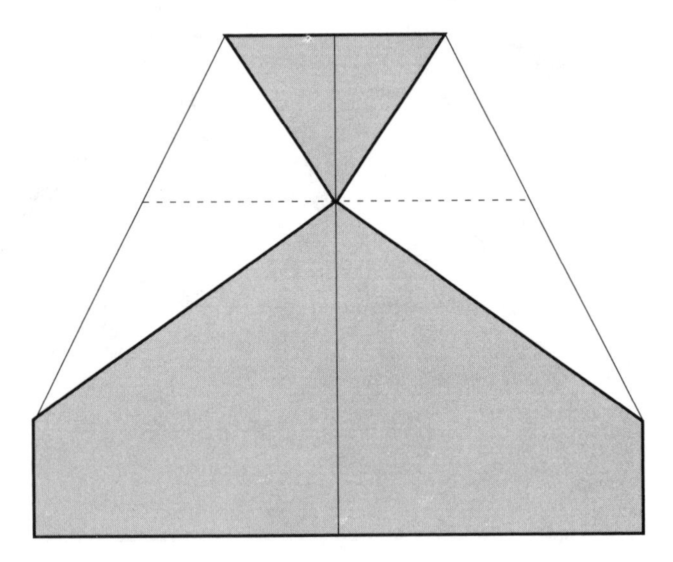

4. Shows the result of Step 3. Fold down the top edge, carrying the fold through the point shown. Be certain you keep the central crease in alignment.

5. Now make folds on both sides so that point A touches point B on each side.

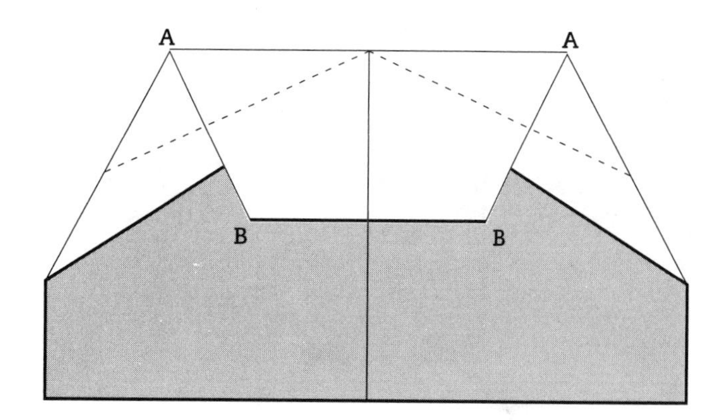

6. Shows the result of Step 5. Now fold down the tip so that it touches the horizontal edge shown.

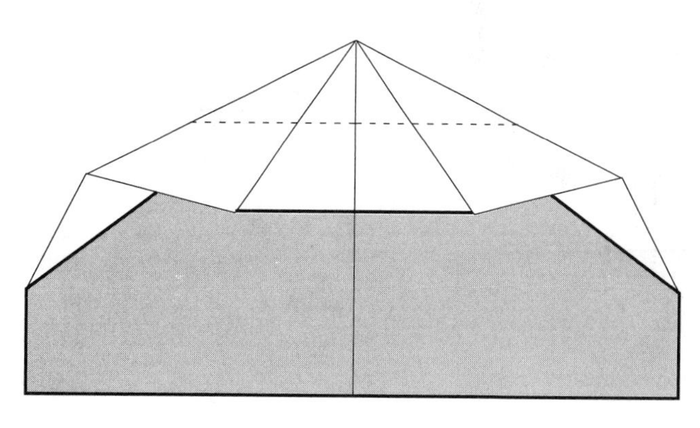

7. Shows the result. Turn the paper over.

Fold in half along the central crease, aligning the wings.

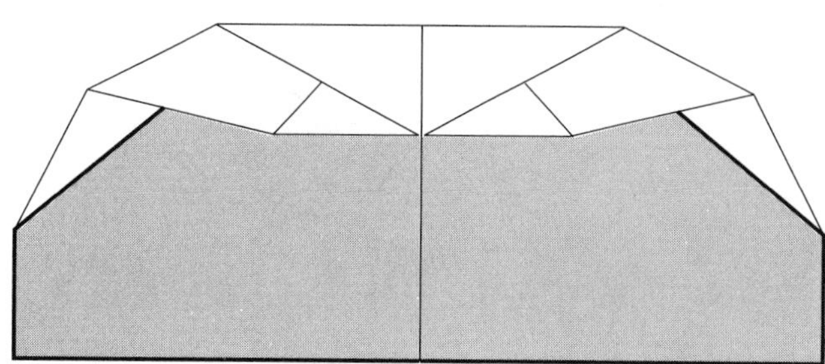

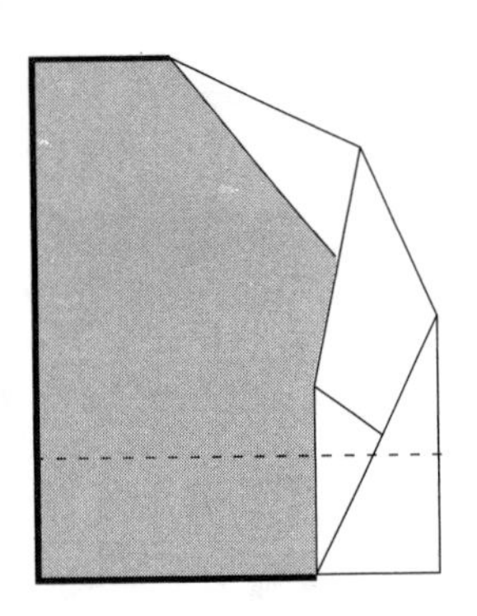

8. Fold each wing down approximately 1 inch from the central crease, and parallel to it. Both wings should be folded down exactly the same.

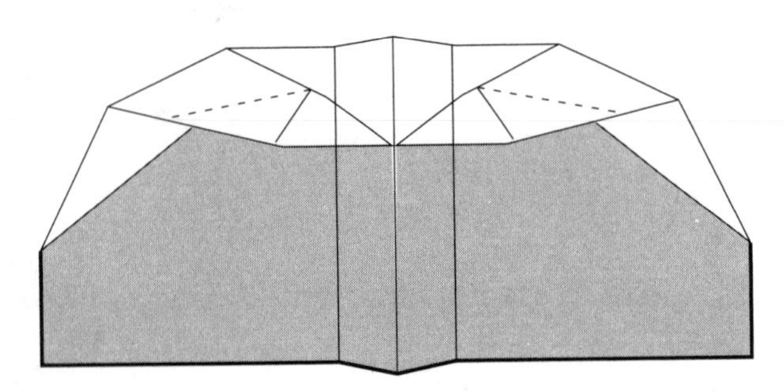

9. Shows the plane from below. Fold flaps up as shown.

Flying instructions for PLANE 4 are on page 16.

Long-winged glider

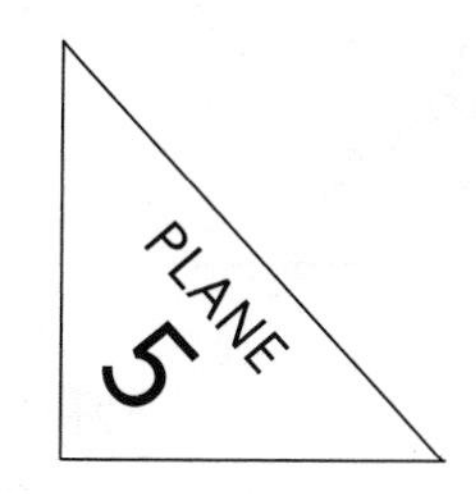

THIS DESIGN is striking because of the length of the wings relative to their width. Because the wing is formed by a diagonal fold, the flat wingspan of this design is a full 13 inches, or 2 inches longer than the longest edge of the paper. That alone makes the plane fun to fold and fly. (Folding instructions are on pages 22–24.)

To fly

Holding the body from underneath, gently toss directly forward. If you have difficulty getting a good toss, try holding the plane from below with your index finger on top in the groove between the wings, pointed forward. Push the plane smoothly forward, and release.

Adjustments

The wings should angle slightly upward from the body when the plane is thrown. The key to a steady flight lies in the wingtips, which you will note bulge downward at the forward edge. You will find that if you crease this region flat, the plane will dive into the ground; if it bulges too much, the flight will be boggy and unsatisfying, and so it should be only partially flattened. If the plane veers to one side, try flattening the forward edge of the wingtip slightly on the same side as the veering direction to straighten the course.

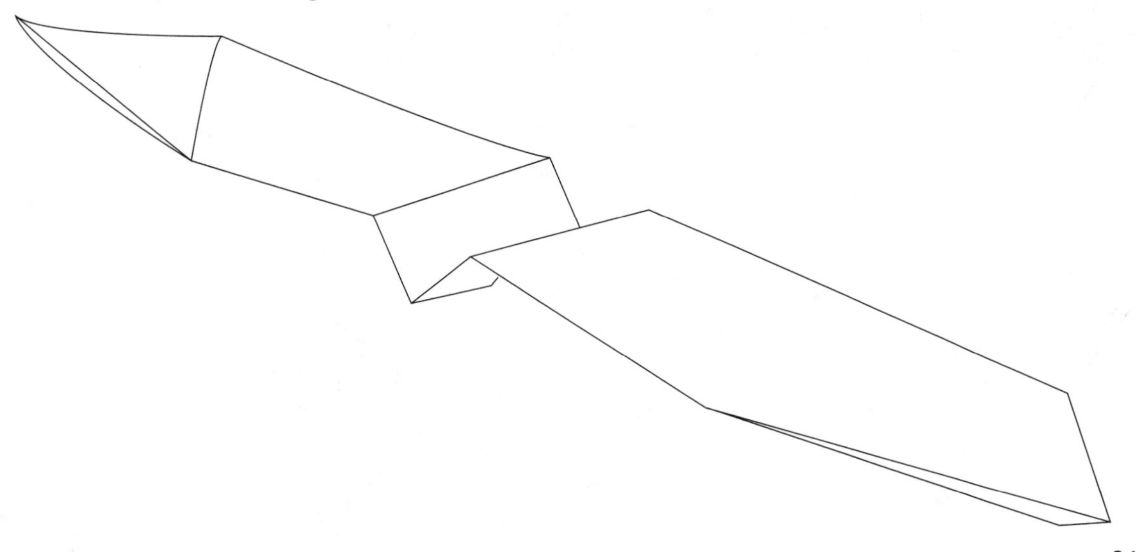

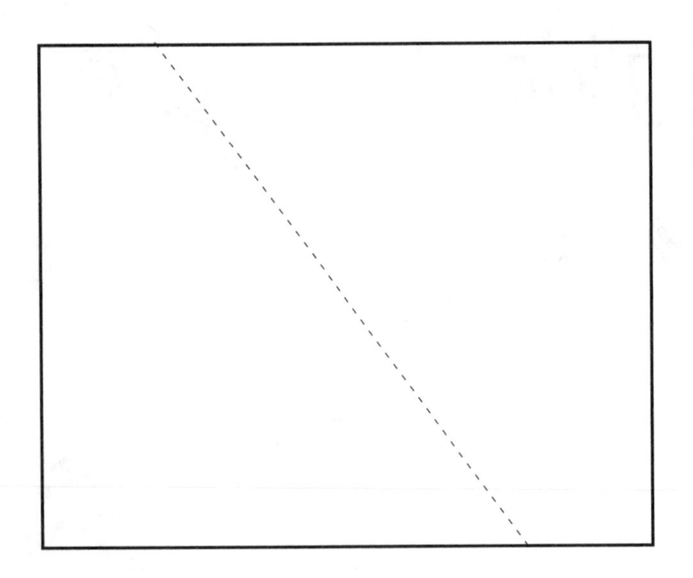

1. Make a diagonal fold as shown by folding one corner of the paper to the corner exactly opposite. Place the paper as shown in Fig. 2.

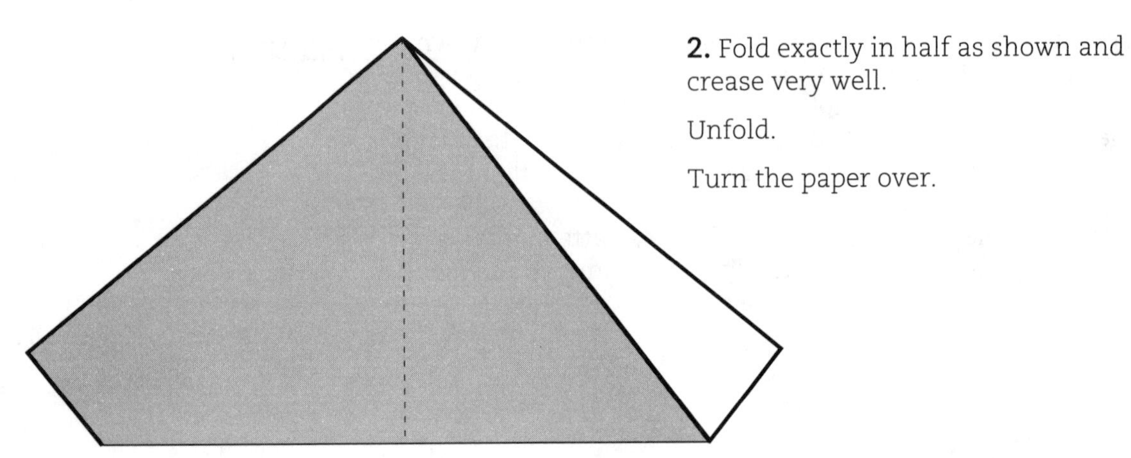

2. Fold exactly in half as shown and crease very well.

Unfold.

Turn the paper over.

3. Fold the tip down to the bottom edge, keeping the central crease aligned.

Unfold.

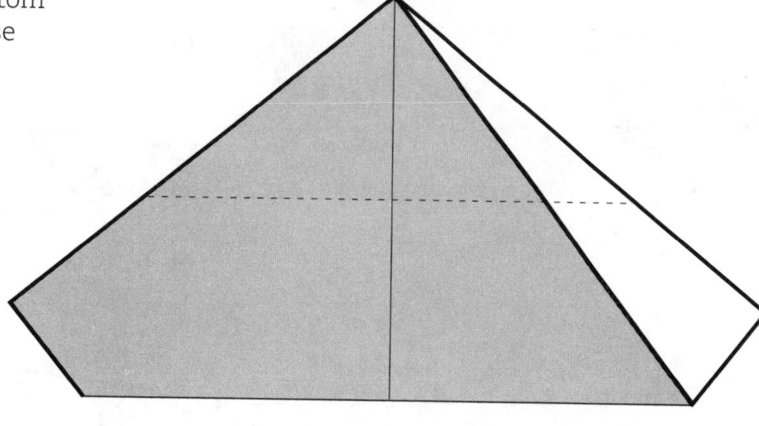

4. Fold the tip down to the horizontal crease you have just made.

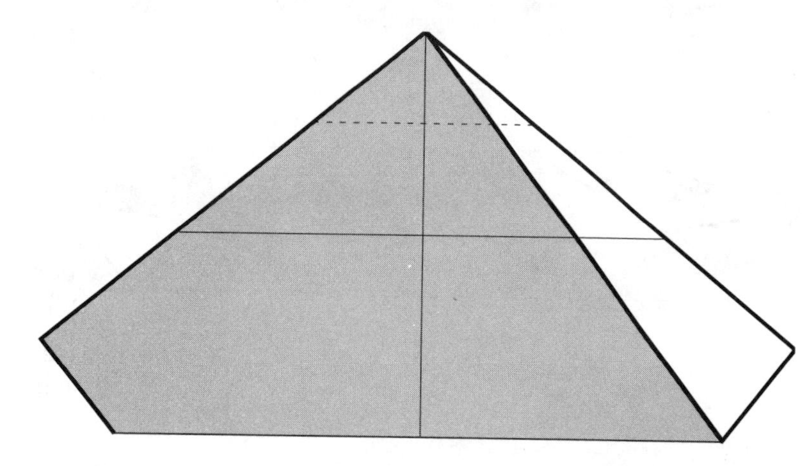

5. Fold the top edge down to the horizontal crease.

Fold again along the horizontal crease, as shown.

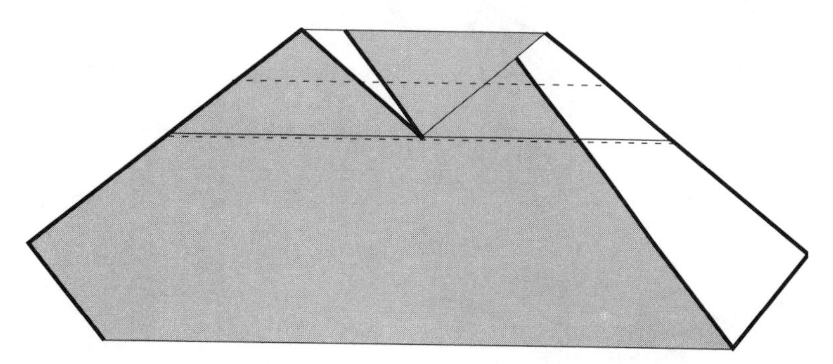

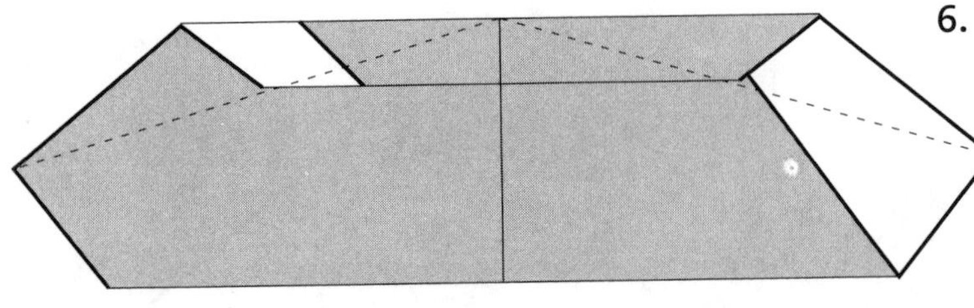

6. Carefully make a diagonal fold on each side from the front of the central crease to each wingtip, as shown. The result should look like Fig. 7.

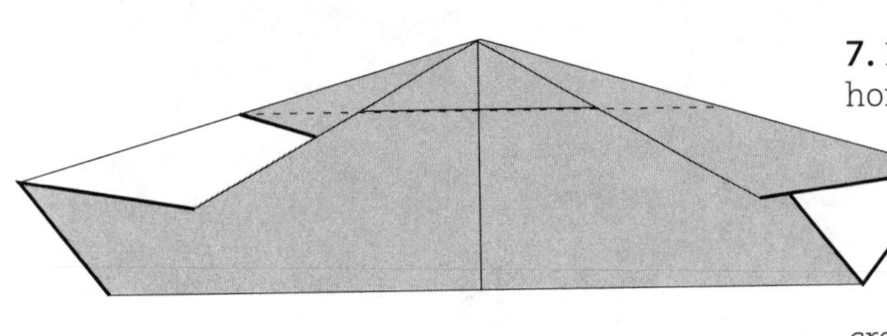

7. Fold the tip down along the horizontal folded edge as shown, being careful to keep the central crease lined up. This will cause something of a bulge in the forward edge of each wingtip; that is OK. Do not crease down this bulge, as it is important to stable flight.

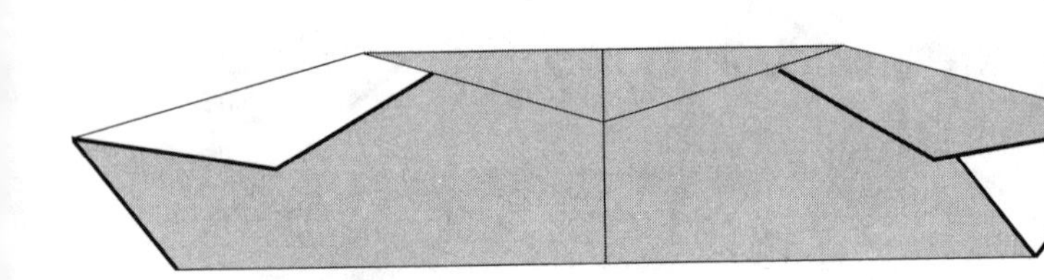

8. Shows the result of Step 7. Turn the plane over.

Fold in half along the central crease, making certain that the wings match each other exactly.

Crease the back edge (the thin edge) of each wing to flatten, creasing from the center outward.

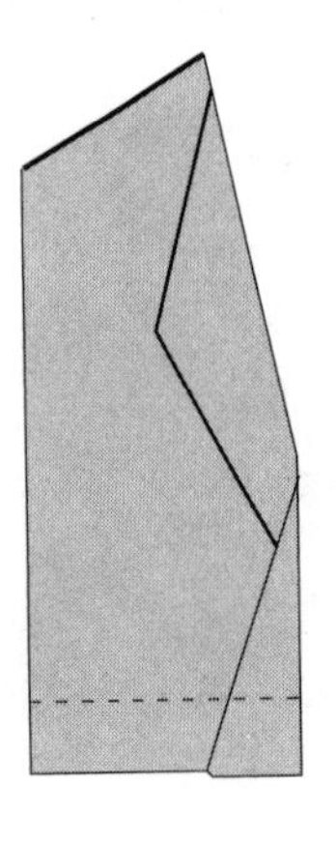

9. Fold down each wing about ¾ inch from the central fold, as shown. Be certain to fold both wings exactly alike.

Flying instructions for PLANE 5 are on page 21.

Finned wings

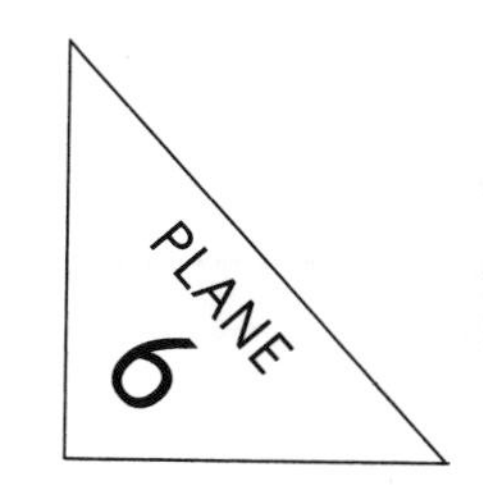

ANOTHER VARIATION on the straight-winged theme, the folded fins at the end of the wings of this plane add good looks and stability. (Folding instructions are on pages 26–30.)

To fly

Hold from beneath and toss directly forward.

Adjustment

If the fins seem to bulge outward, push in at the side and bend the top and bottom fins together, then restraighten. Experiment with the angle of the fins to steer the plane. If lift is poor, you can slide your finger forward along the underside of the wing, lifting it slightly from the fold of paper underneath the forward edge.

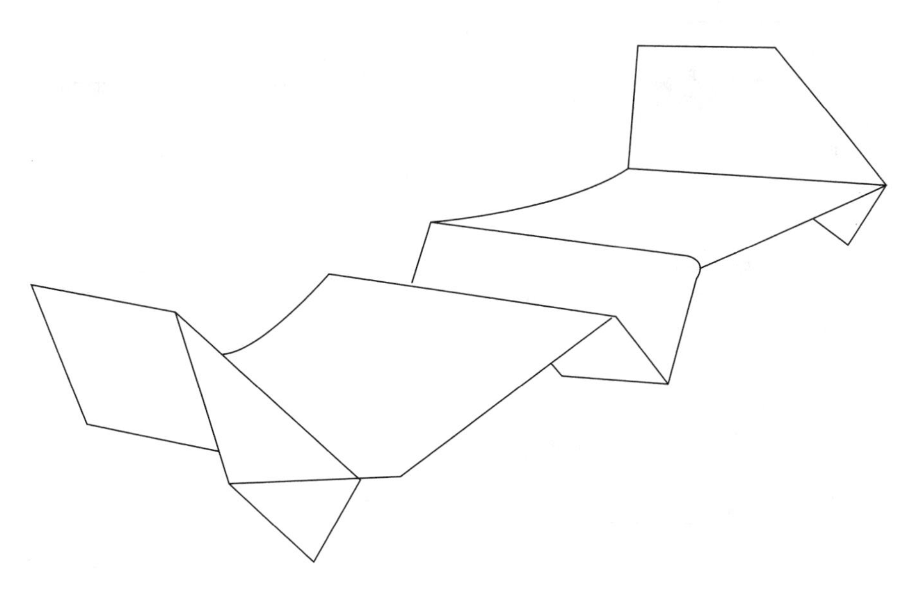

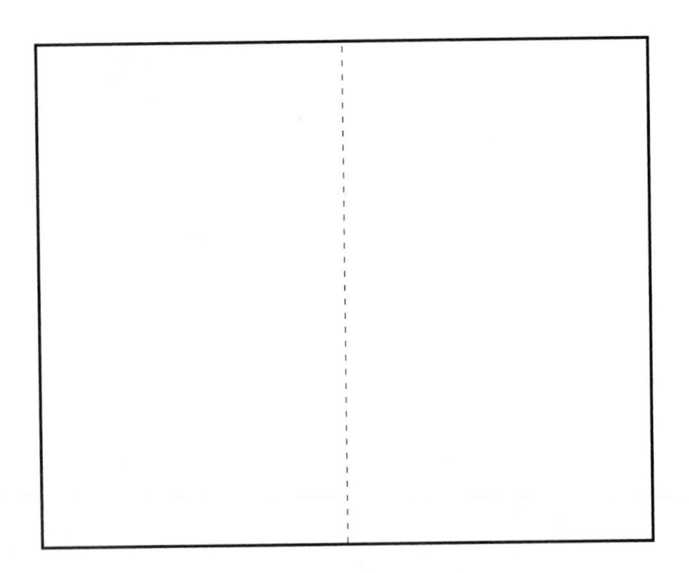

1. Make a vertical central crease as shown, by aligning the sides of the paper. Crease well.

Unfold.

Turn the paper over.

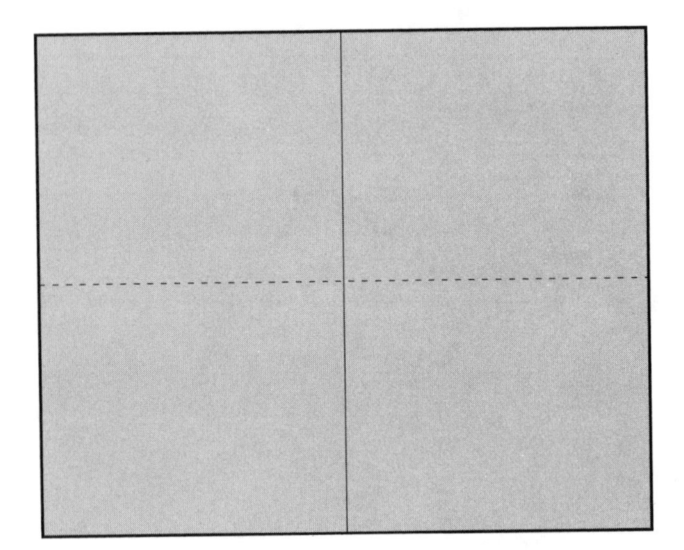

2. Make a horizontal crease as shown by aligning the top edge with the bottom. Crease, keeping the vertical crease lined up.

Unfold.

3. Fold down each top corner so each corner rests against the horizontal crease, as shown.

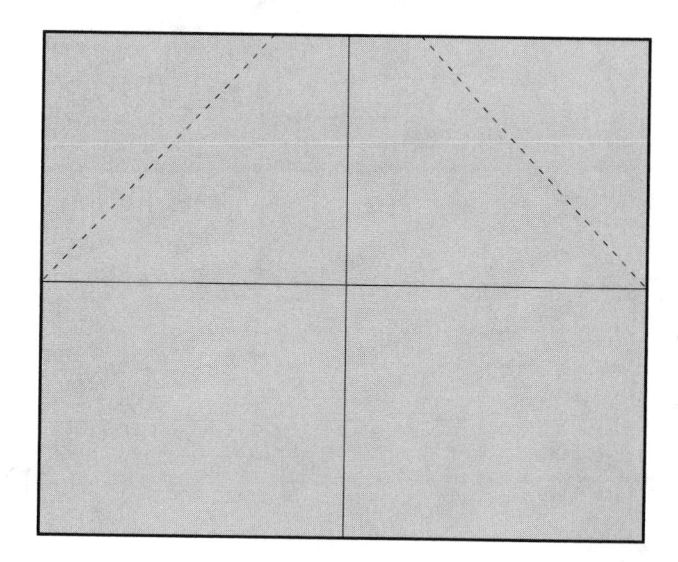

4. Make another similar fold on the right side, aligning the diagonal edge with the horizontal crease.

Unfold.

Repeat the fold on the other side.

Unfold. If you did this carefully, the diagonal folds should cross in the exact center.

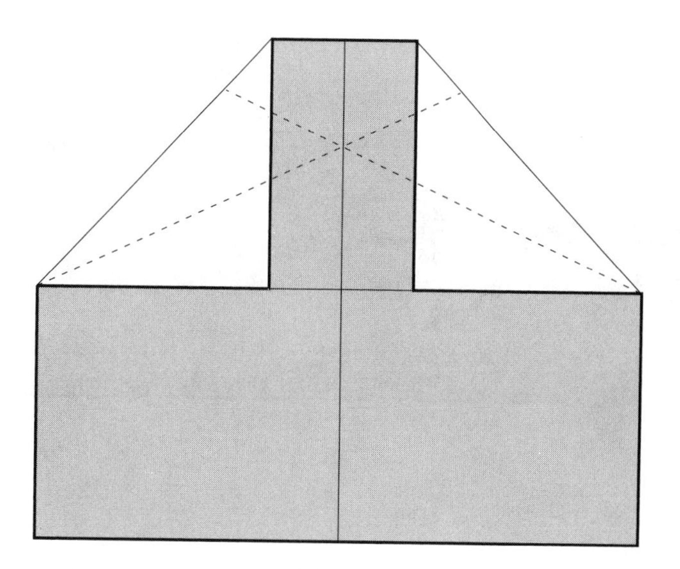

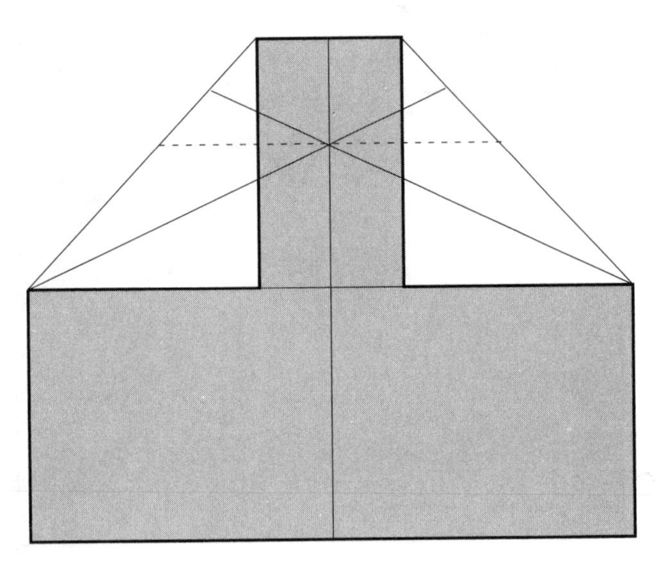

5. Make a horizontal fold exactly through the intersection of the two creases you made in Step 4, as shown. Be certain you keep the central crease aligned. Crease well.

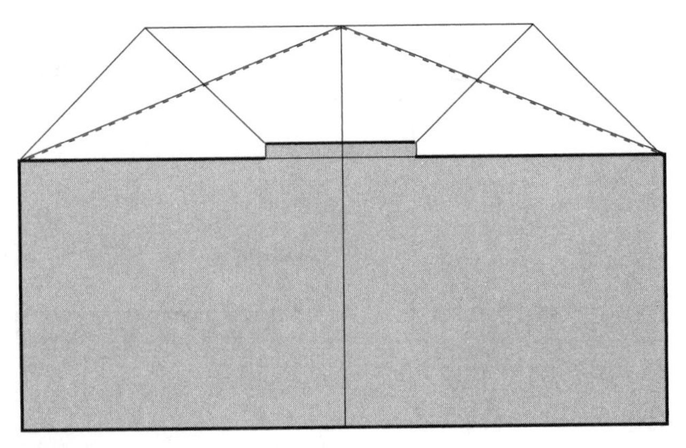

6. Fold down along the diagonal crease on each side, as shown. Crease well.

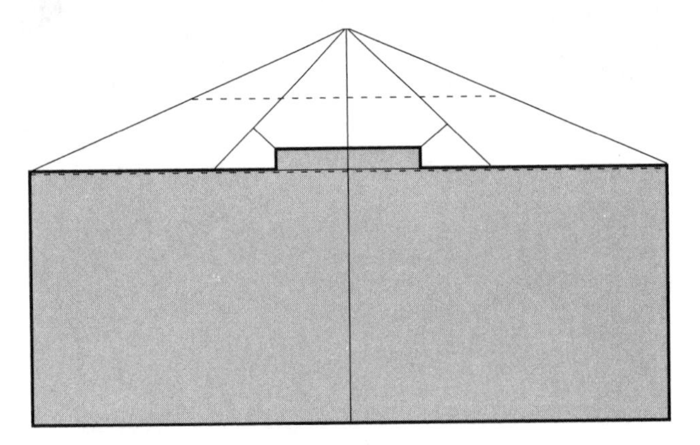

7. Fold the tip down until it just touches the horizontal crease in the exact center. Crease well.

Fold again along the horizontal crease, as shown.

8. Fold down again along the edge shown, keeping the central crease aligned.

Unfold.

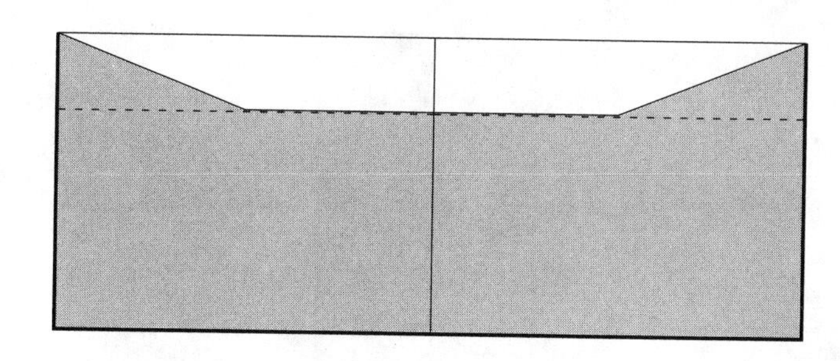

9. Fold down the corners on each side against the horizontal crease you just formed.

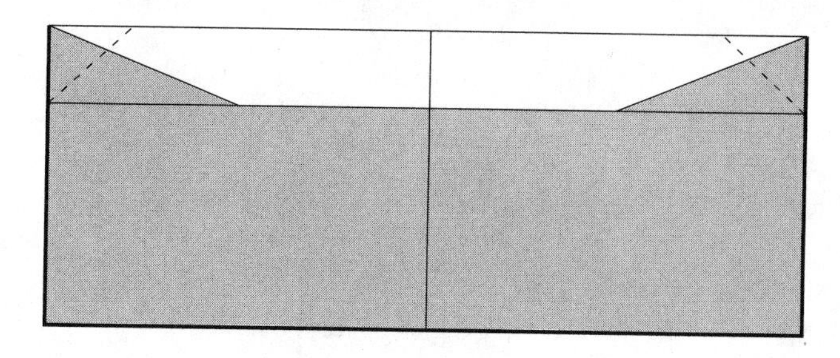

10. Fold down along the horizontal crease.

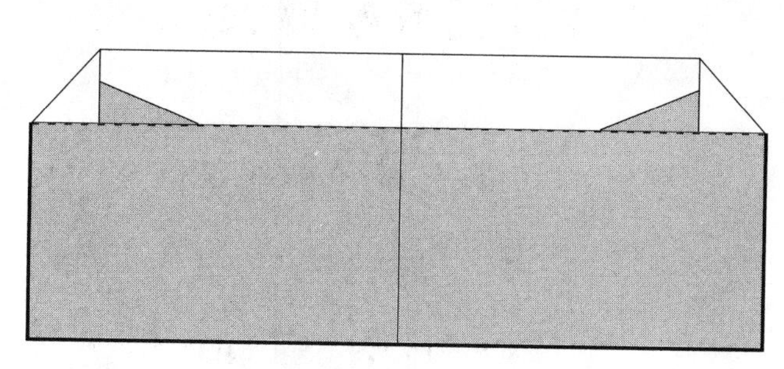

11. A fin should be folded at the end of each wing in the location shown, but this fold must be made away from the side shown. Turn the plane over, and make this fold.

Fold the plane in half along the central crease.

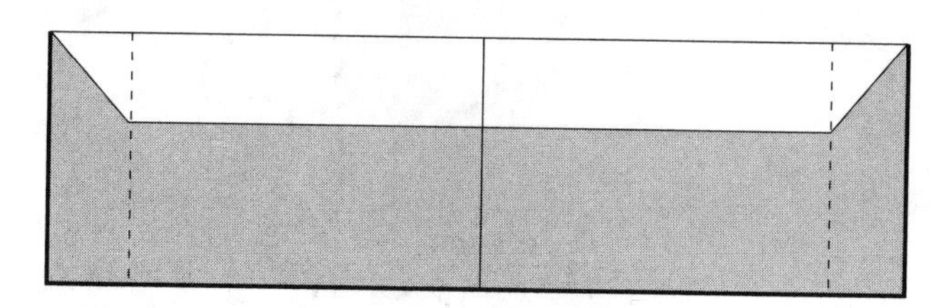

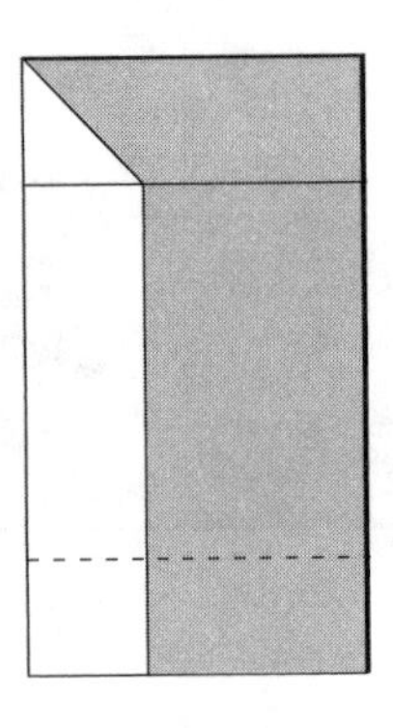

12. Fold down each wing approximately 1 inch from the body, and parallel to the body. Both wings should be folded exactly the same.

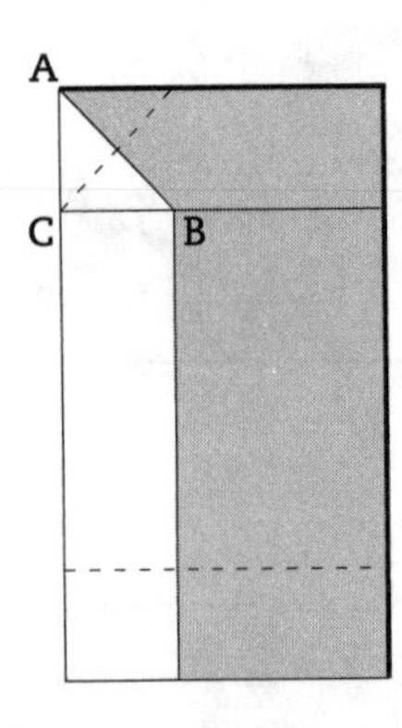

13. Now make an accordion fold on the front edge of the wing fins. Slide your finger into the pocket formed by edge AB. Bring corner A to point B, forming both an upper and a lower fin. Crease both the upper and lower side of the fin so that they appear as in Fig. 14, straightening out fold AC in the process.

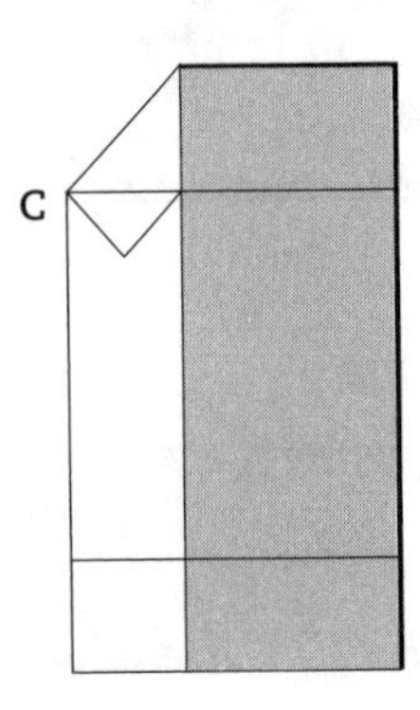

14. Shows the result of Step 13. Adjust the plane so the wings stick straight out and the fins are aligned vertically at the ends of the wings.

Flying instructions for PLANE 6 are on page 25.

Aerobat

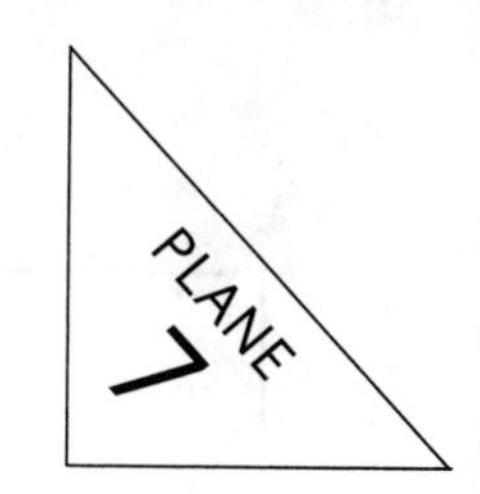

THIS LITTLE PLANE closely resembles a playground classic I used to fold as a boy, with a bit more refinement. It flies well indoors or out, and with a stiff throw will perform loops before settling down for a long flight. (Folding instructions are on pages 32–35.)

To fly

Hold the plane from beneath and toss directly forward. Outdoors, this plane can be thrown hard for long flights if there is not too much wind. In this case, throw the plane strongly on a slightly upward path into the wind, or try throwing almost directly upward.

Adjustments

Make certain the final wingtip folds are equal. If the plane needs more lift, bend the back edge of each wing slightly upward. This will also help the plane perform loops if thrown strongly. If the plane climbs too much or loops too tightly, straighten out the back edge, or bend the back edge slightly downward.

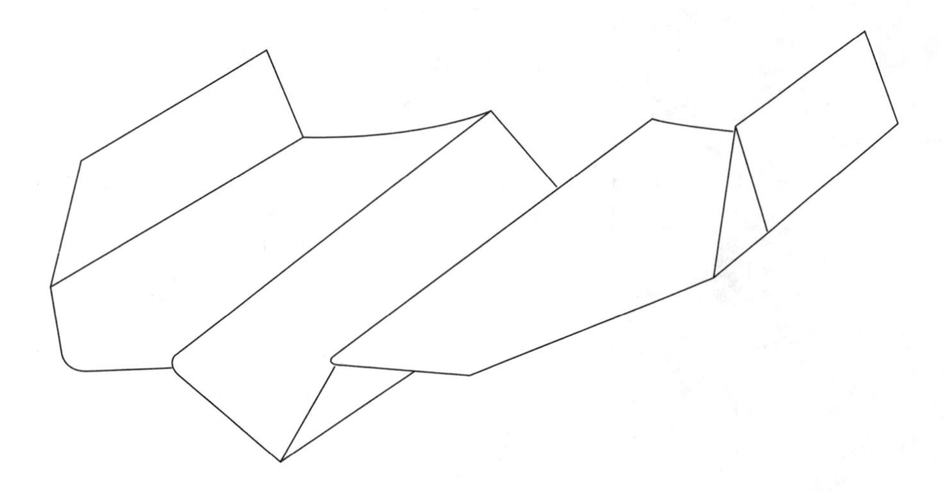

1. Placing the paper as shown, make a fold extending from the top left corner, by aligning the top edge with the left vertical edge. Crease well.

Unfold.

Repeat the same fold with the opposite top corner.

Crease, and unfold.

Turn the paper over.

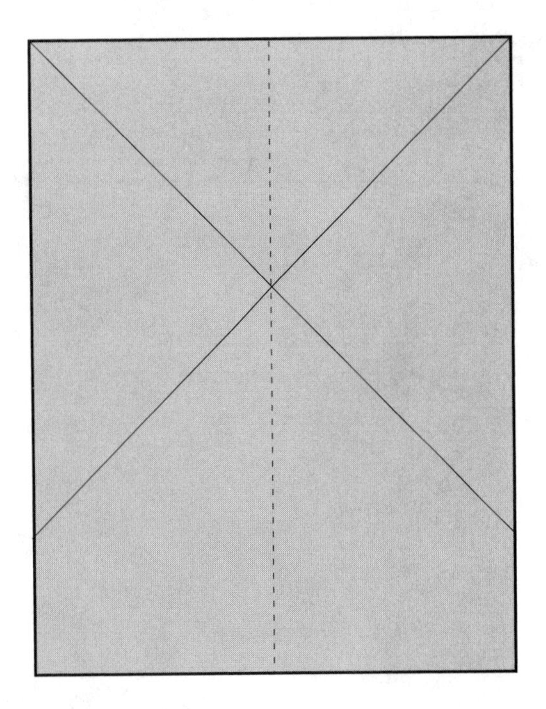

2. Make a central crease vertically as shown by aligning the right side of the paper to the left.

Unfold.

3. Make a horizontal crease as shown, so that this crease passes through the intersection of all the creases already made. Keep the central crease carefully aligned.

Unfold. The result should look like Fig. 4.

Turn the paper over again.

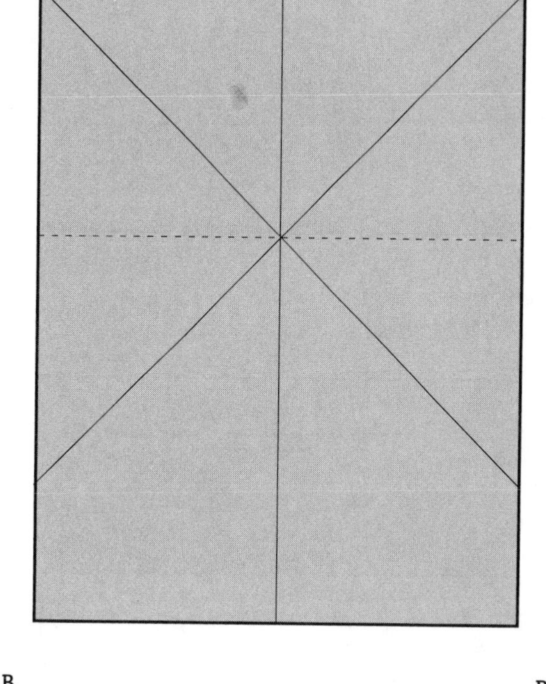

4. Now make an accordion fold as follows: Push in with your finger on the point where all the lines intersect. This should make the center bulge away from you. Grasp point A on each side of the paper and draw each side downward and toward the center of the paper until they touch the central crease. Hold them there. Now bring down the top corners labeled B so that they rest against point C on each side. When you have completed the fold successfully, recrease all edges. The result should look like Fig. 5.

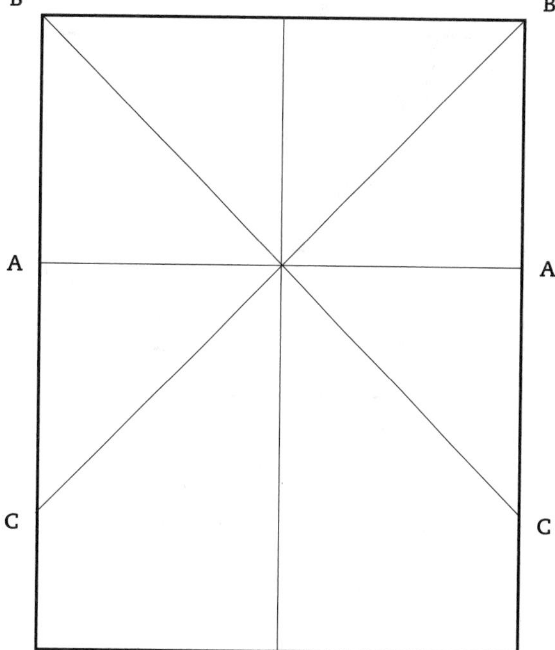

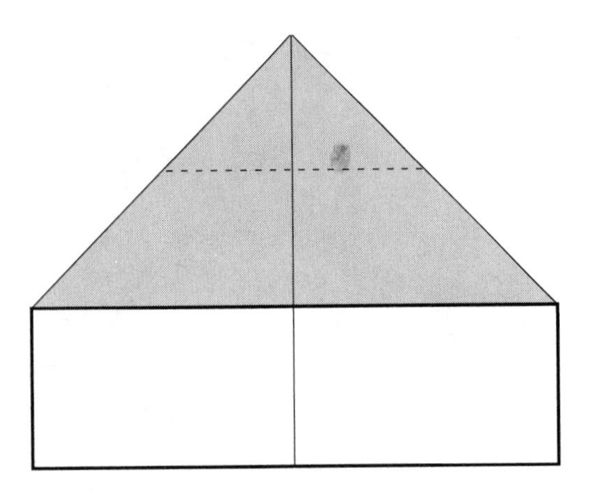

5. Fold the tip down so it rests against the horizontal paper edge in the exact center, as shown. Note that this fold creates a triangle from the top edge, and that little pockets are formed in the sides of the triangle. Open these pockets gently with your finger.

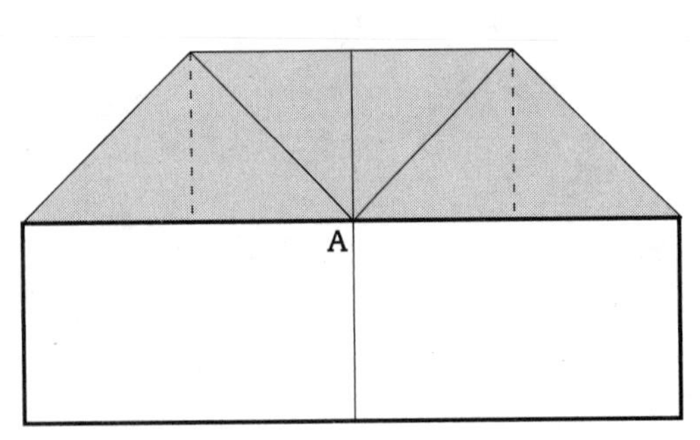

6. On each side you should have two wings: a small, pointed upper wing resting on a larger lower wing. On each side, fold the pointed upper wing inward so that the point just touches the tip of the triangle formed in Step 5, labeled A.

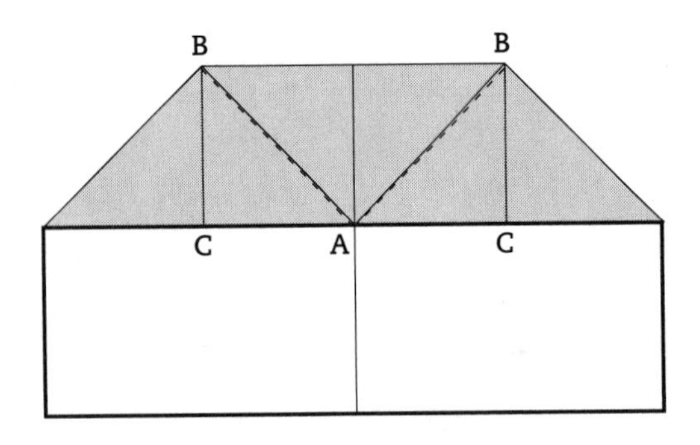

7. Now fold the small wing upward diagonally on each side as shown. Partially unfold. Tuck tip C on each side into the little pocket you formed in Step 5, labeled here AB.

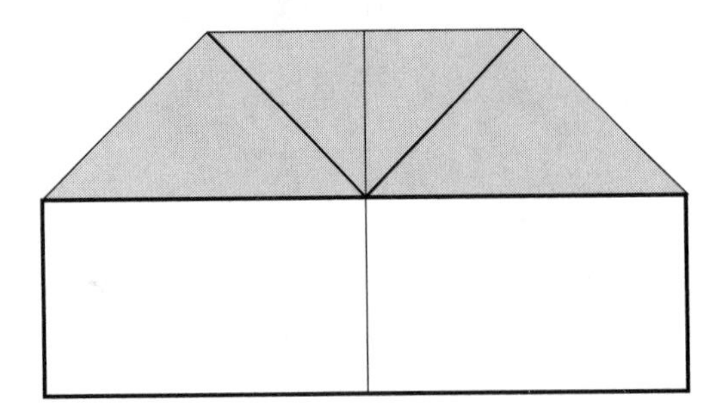

8. Shows the result of Step 7. Turn the plane over.

Fold in half along the central crease.

9. Fold down each wing, so that point A exactly matches point B on each side. Be certain to fold both wings exactly the same.

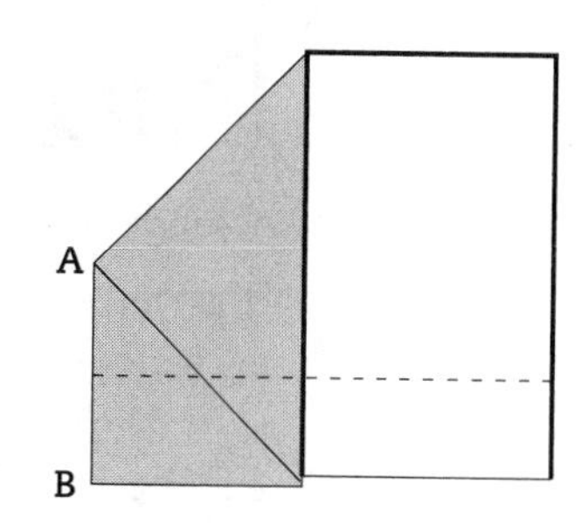

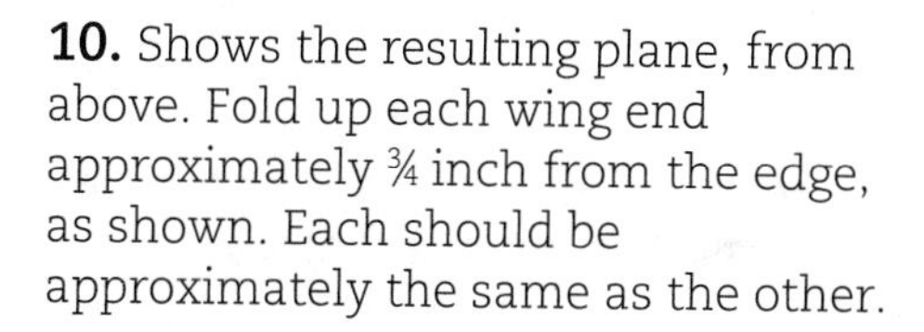

10. Shows the resulting plane, from above. Fold up each wing end approximately ¾ inch from the edge, as shown. Each should be approximately the same as the other.

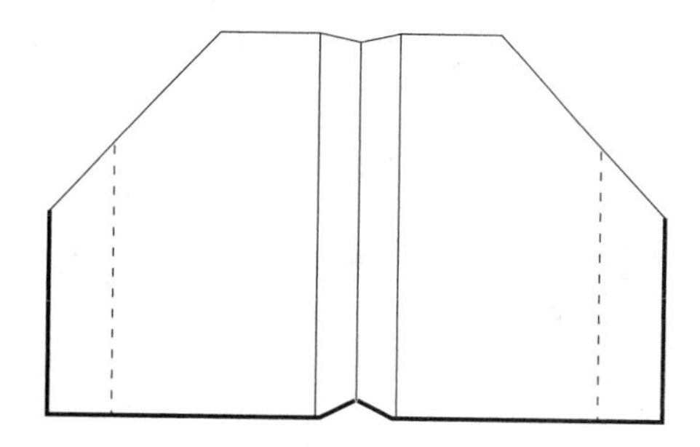

Flying instructions for PLANE 7 are on page 31.

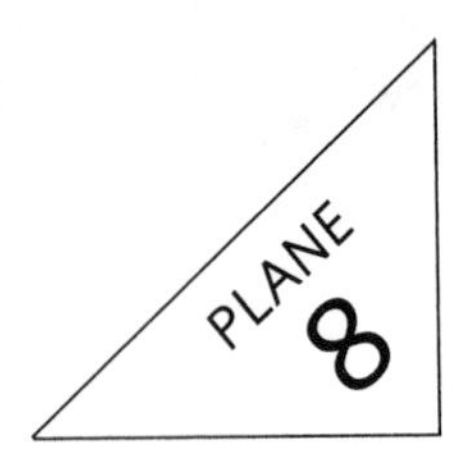

Midget

THE FIRST STEP of this design is to fold the paper in half; the entire plane is folded from the resulting half-sheet. The midget is, therefore, both compact and quick, but still with an excellent glide. (Folding instructions are on pages 37–40.)

To fly

Hold from below, and throw firmly forward.

Adjustments

If the plane does not fly well, recrease the wings and check for symmetry. This plane sometimes needs some elevator for extra lift. Bend up the back edge of both wings slightly.

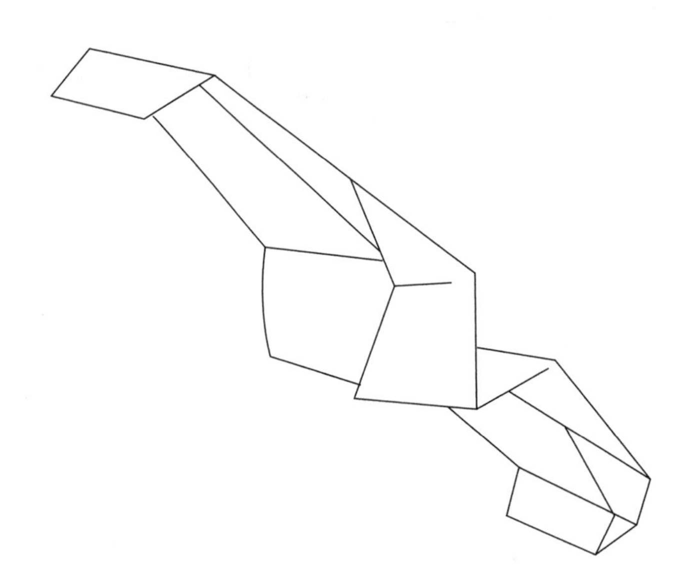

1. Fold the paper in half, aligning the bottom edge with the top as shown. Do not unfold.

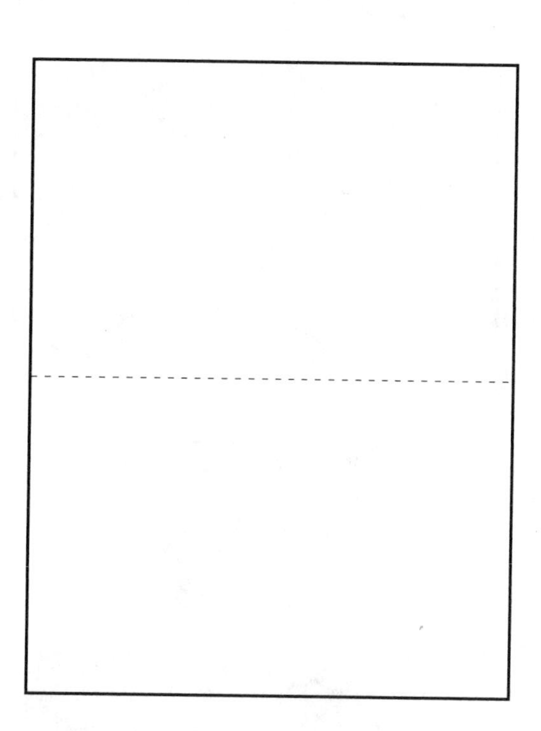

2. Fold the paper in half again, as shown.

Unfold.

Turn the paper over.

3. Making sure that the folded edge is downward, fold down each of the top corners as shown, so that the top edge aligns with the central crease.

Unfold.

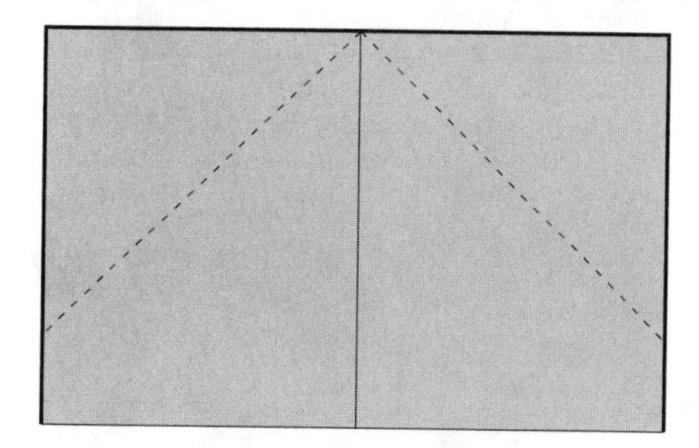

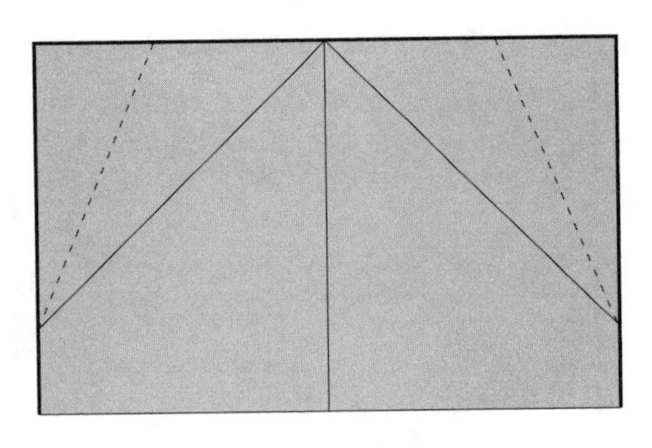

4. Now fold down each of the upper corners, so that the right edge folds against the diagonal crease you just made on each side, as shown.

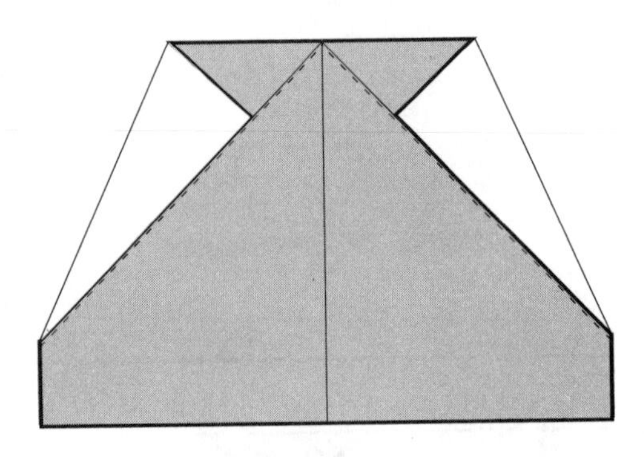

5. Shows the result of Fig. 4. Now fold down each side along the diagonal crease, as shown.

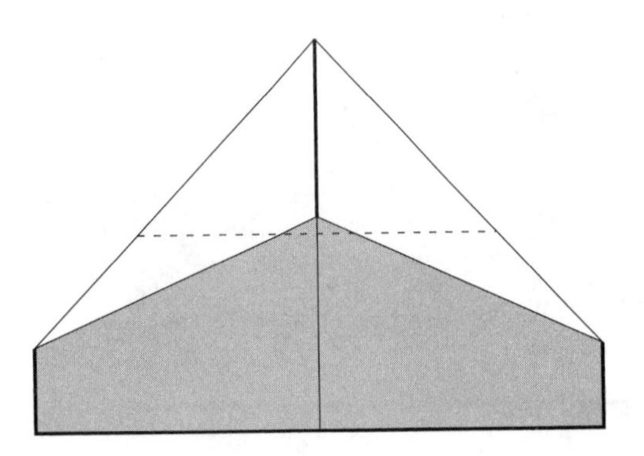

6. Fold the tip of the plane back so it just touches the lower edge of the paper in the exact center.
Crease well, then unfold.

7. Fold down the tip so it just touches the horizontal crease you just made, in the exact center.

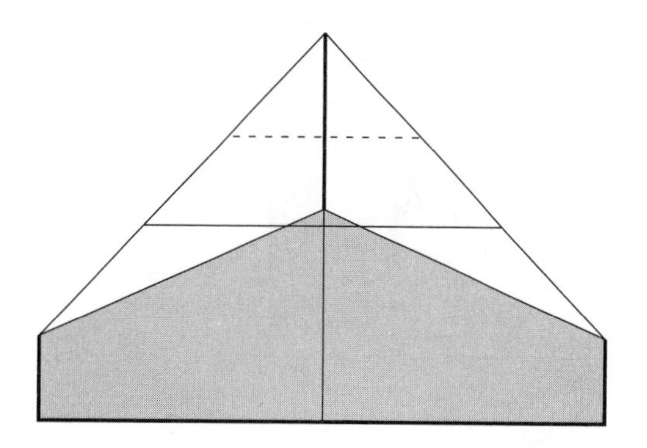

8. Now make a diagonal fold as shown on each side. One end of the fold should be at the front of the plane in the exact center, and the other should touch the outer end of the horizontal crease.

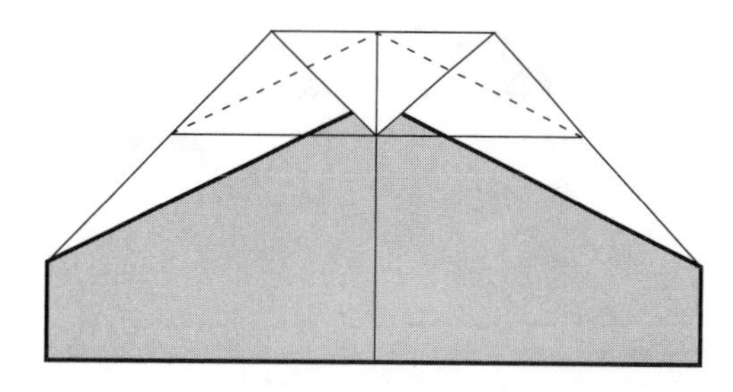

9. This shows the result of Step 8. Now fold down along the horizontal crease as shown.

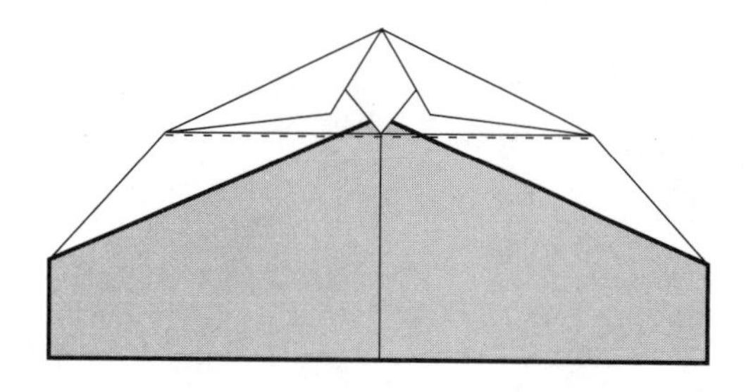

10. The plane should look like this from below. Turn the plane over.

Fold in half along the central crease. The wings should match each other exactly.

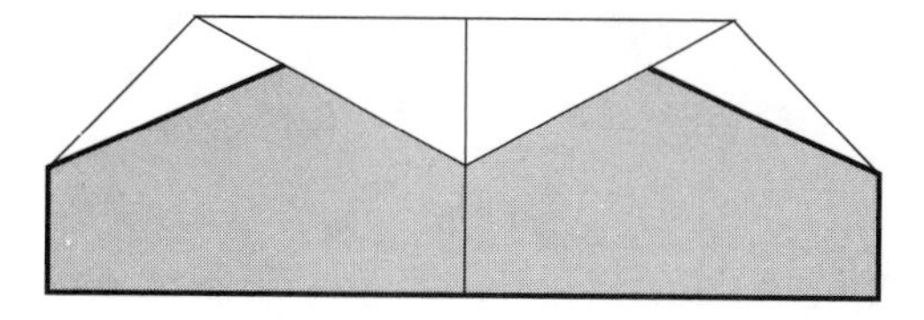

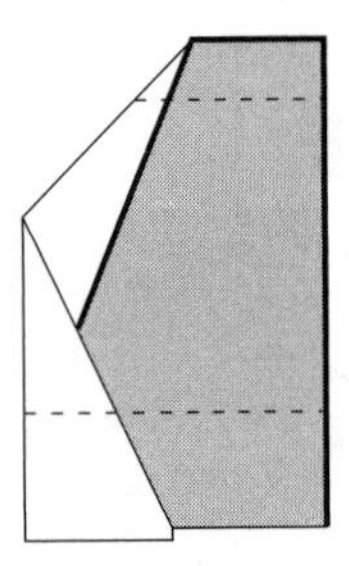

11. Fold down the wings about one inch from the central fold, and parallel to it.

Fold the wingtips down also as shown, approximately ½ inch from the end. Carefully recrease the back edge of both wings.

Flying instructions for PLANE 8 are on page 36.

Fuselage plane

PLANE 9

FOR MOST OF US, the idea of a "real" airplane conjures up the image of a long central body (the fuselage) with a pair of wings in front and a tail behind. While this has been a practical and prolific design in aircraft production, in the world of folded paper airplanes it is a challenge to reproduce. This model meets that challenge, and while it is a little cumbersome to fold (especially with thick paper), it flies very well, and demonstrates some of the features of its real-life counterparts. If you are just starting out, however, try some of the other designs first. (Folding instructions are on pages 42–49.)

To fly

Hold from beneath near the tail, and throw forward forcefully, taking care to throw in a straight line with the nose of the plane.

Adjustments

This is a difficult plane to fold, and is bulky because of the thickness of the many folds. As with all designs, see that the folds and thicknesses are as evenly matched as possible on both sides. This plane may be "steered" much like a real aircraft. Generally, adjustments for turning are accomplished by bending the rear edge of the wings (ailerons) while adjustments for climbing or diving are made by adjusting the tail (elevators). Once you get the plane flying well, you can experiment with quite a number of different effects.

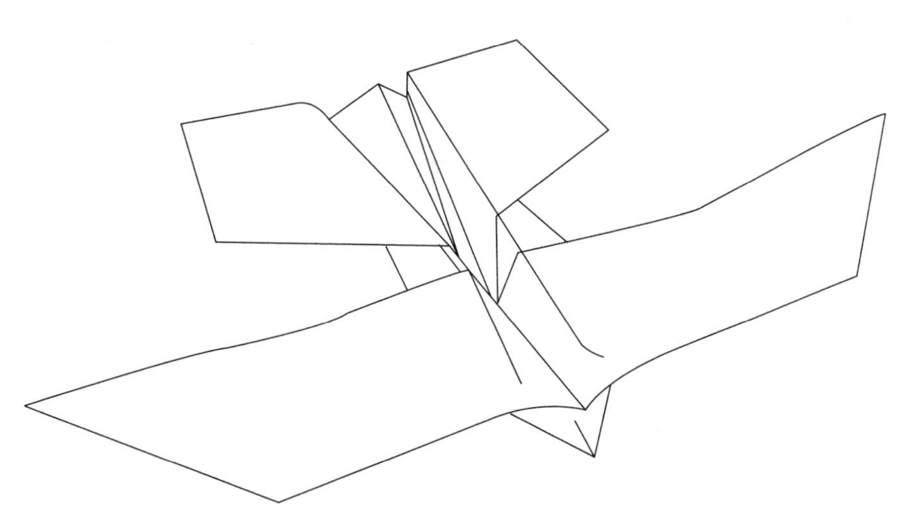

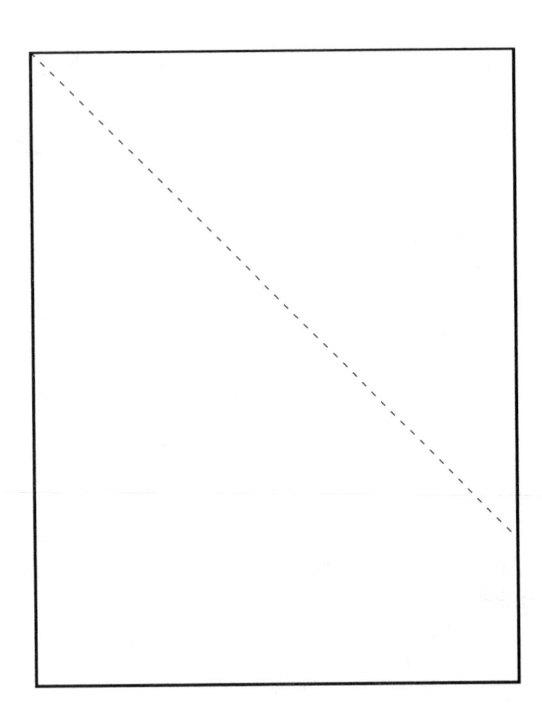

1. Make a diagonal fold from the left upper corner of the paper as shown, by aligning the top edge of the paper with the vertical left edge. Crease well.

Unfold.

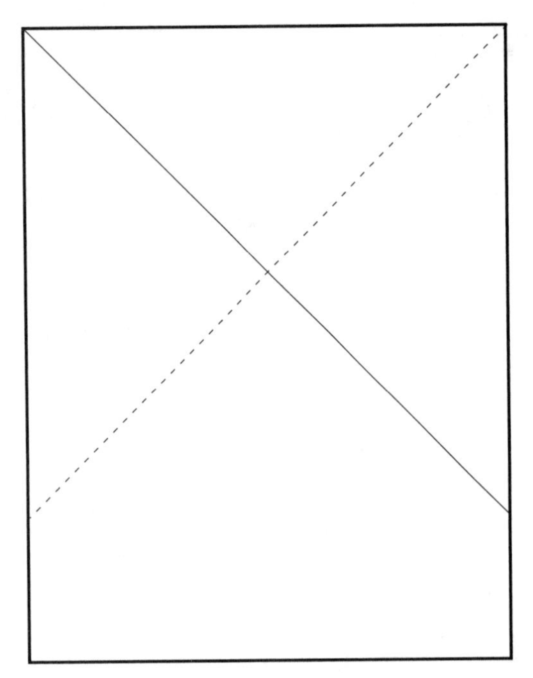

2. Make a similar diagonal fold extending from the right top corner. Crease well.

Unfold.

Turn the paper over.

3. Make a vertical midline crease by aligning the left side of the paper with the right.

Unfold.

Make a horizontal crease exactly through the intersection of the previous creases. This is easily done by aligning point A with point B on each side.

Unfold.

Turn the paper back over.

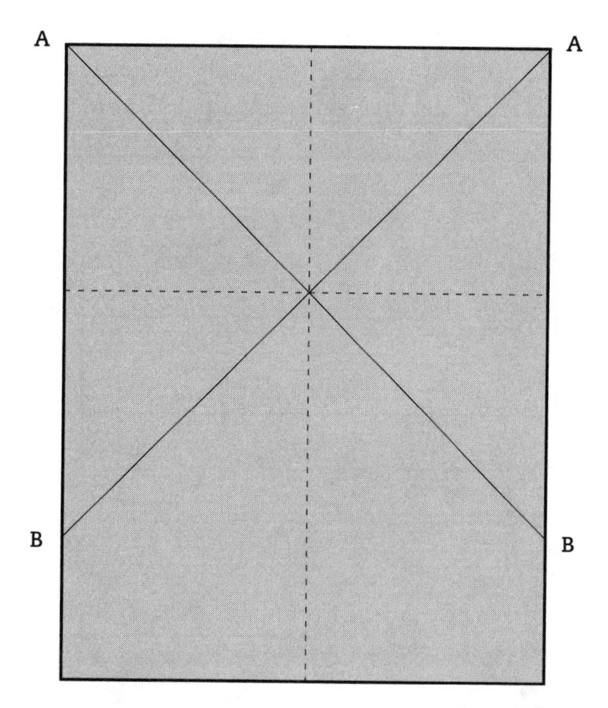

4. Make an accordion fold as follows: Press in on the point where the folds intersect, so that it bulges away from you. Grasp the two ends of the horizontal crease, both marked C, and pull downward and inward. This should start the accordion fold. Fold the corners marked A exactly down on point B on either side. Crease all edges well. The result should look like Fig. 5.

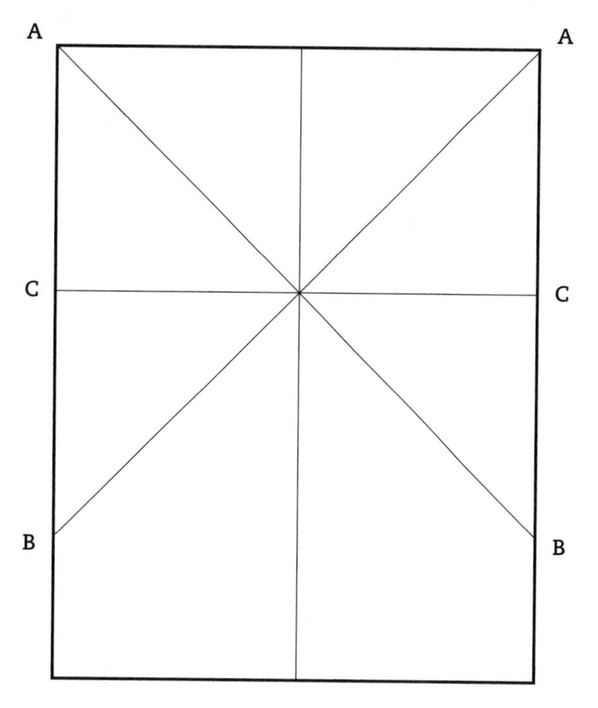

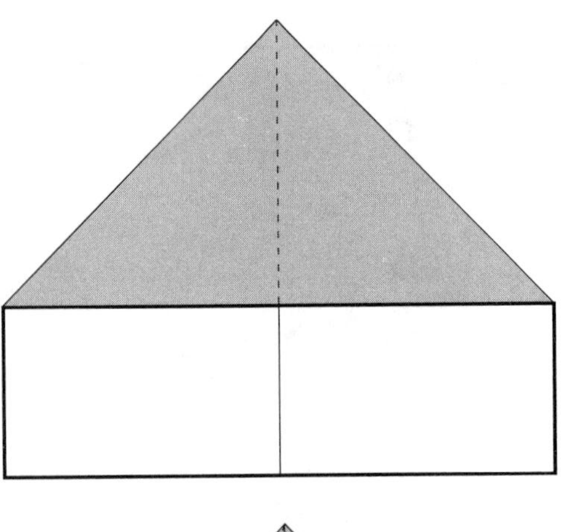

5. Note that you have two sets of wings, a small upper set resting over a larger lower set. Fold the small wing on the left over to the right side.

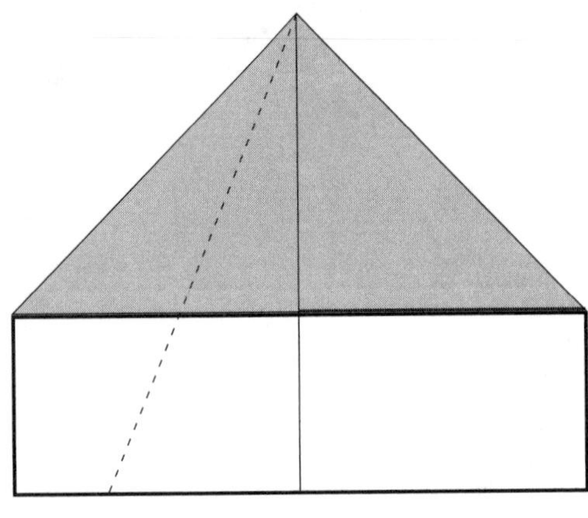

6. Fold down the left side of the plane as shown, so that the diagonal edge aligns exactly with the central crease. The result should look like Fig. 7.

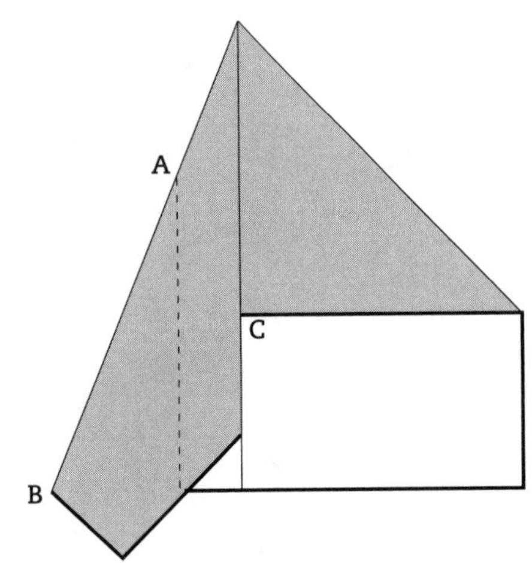

7. Fold the left tail over to the right, so that diagonal edge AB when folded over just touches point C as shown in Fig. 8. Note that the horizontal bottom edge of the paper should line up perfectly.

8. Shows the finished fold of Step 7. Unfold the fold you just made.
Fold both the small wings from the right side to the left.

9. Fold down the diagonal edge on the right side, aligning with the central crease, as you did on the left in Step 6.

10. Fold the right tail over, touching the diagonal edge to point C as shown on the left in Fig. 8.

Unfold.

Fold one of the small wings back to the right. The result should look like Fig. 11.

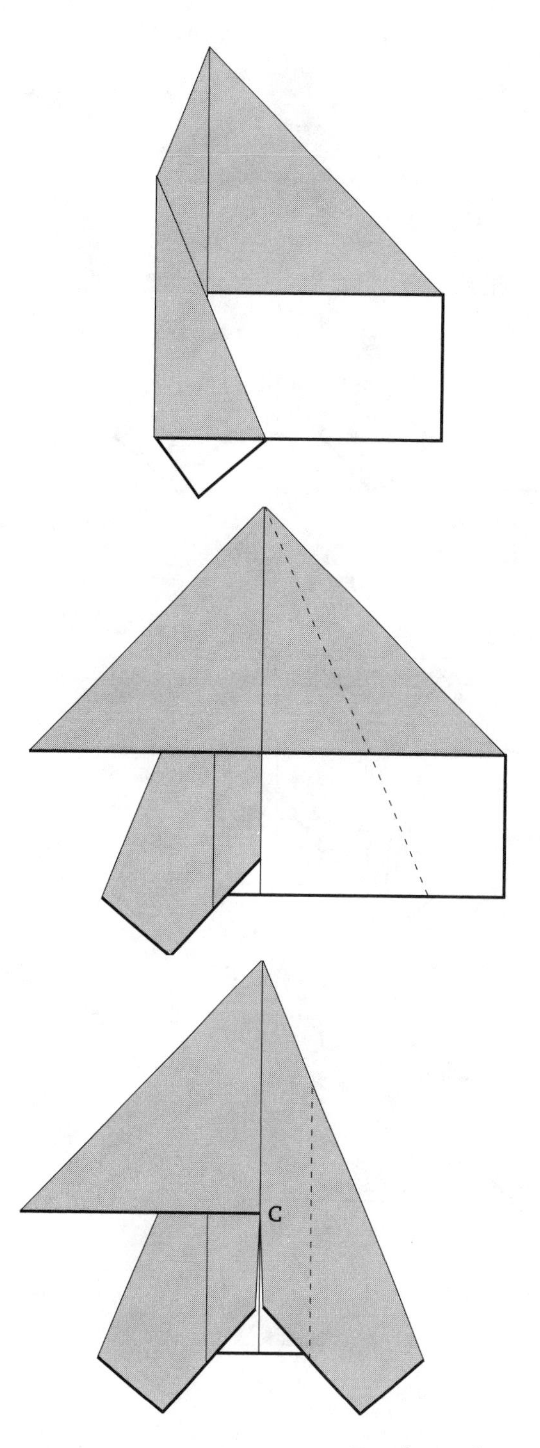

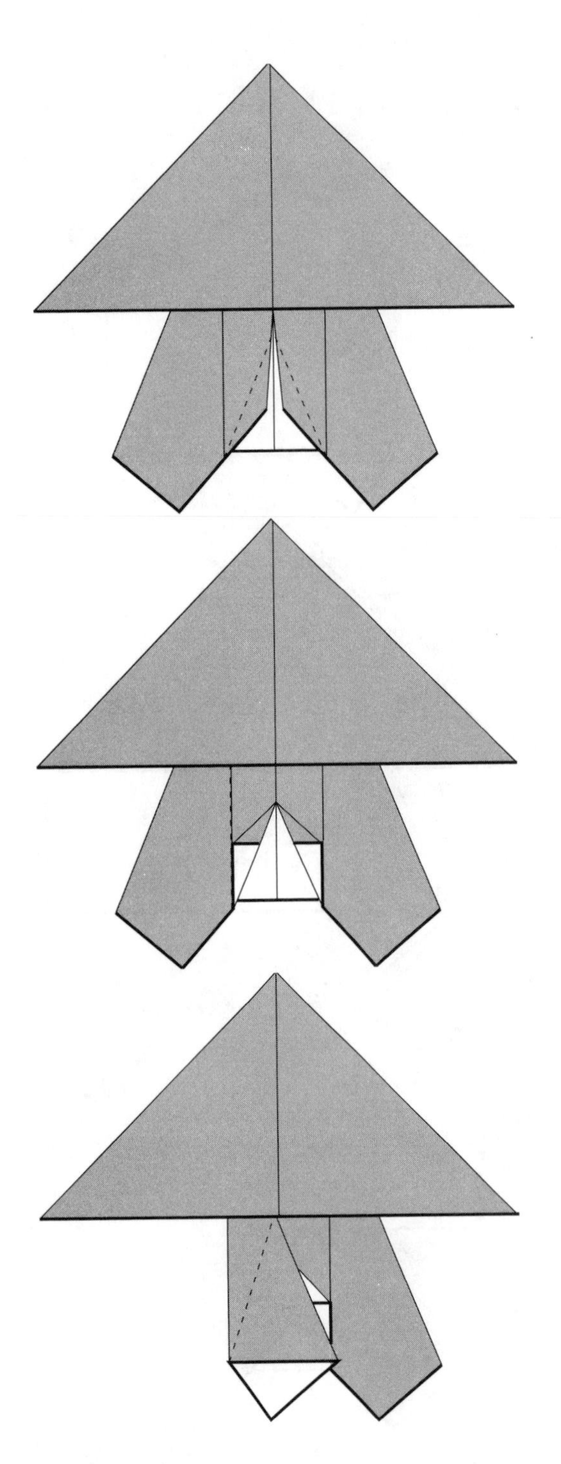

11. Fold up the small corner shown on each side, so it looks like Fig. 12.

12. Now fold in the left tail along the vertical fold created earlier.

13. Fold out the left side of the tail as shown. Your diagonal fold should extend from the center crease on top to the bottom left corner.

14. Shows the result of Step 13. Now fold in the right side of the tail along the vertical crease as shown.

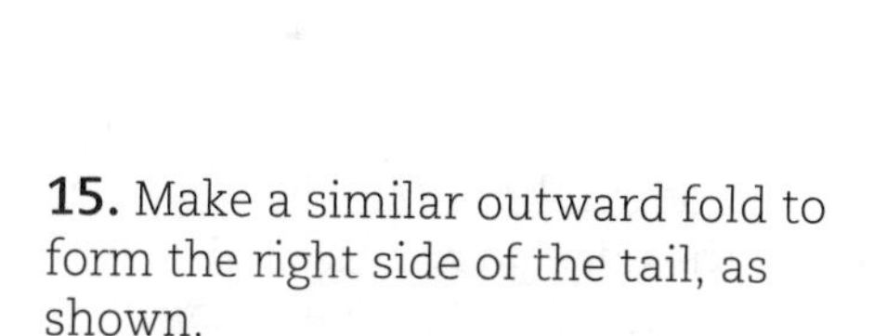

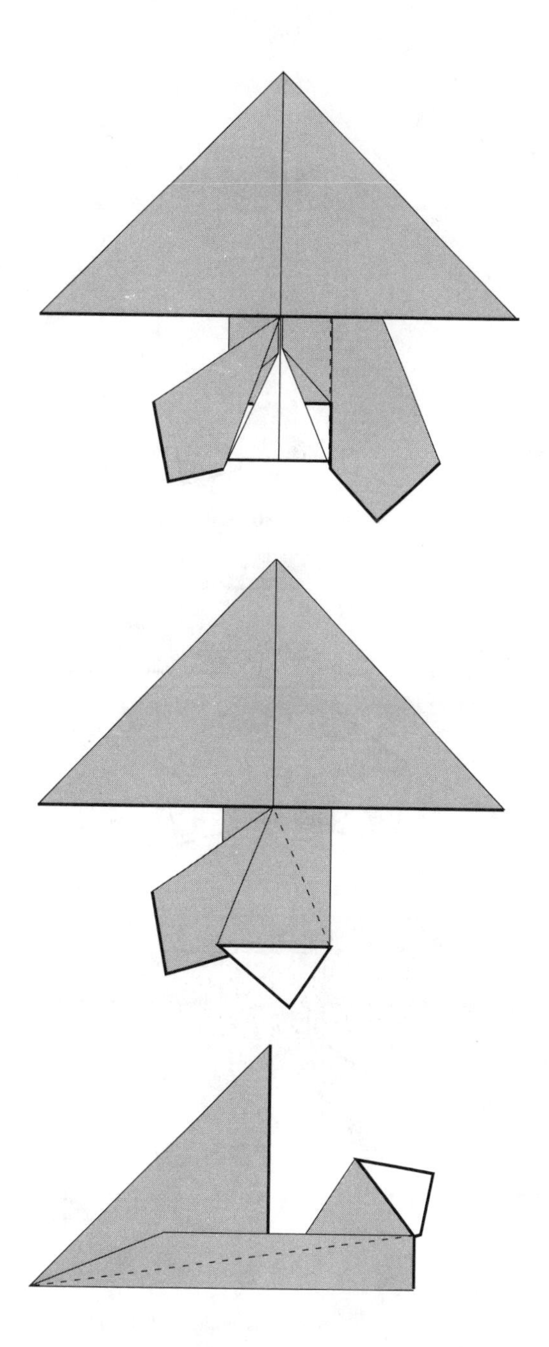

15. Make a similar outward fold to form the right side of the tail, as shown.

Fold the plane in half along the central crease. The wings and tail should match up fairly exactly.

16. Fold down the wings as shown. The fold on each side should extend from the front tip of the plane to the corner next to the tail, as shown in the illustration. The tail on each side should be folded down too.

Make certain the wings and tail match each other exactly, and unfold.

Unfold the central crease, opening the plane out.

Turn the plane over.

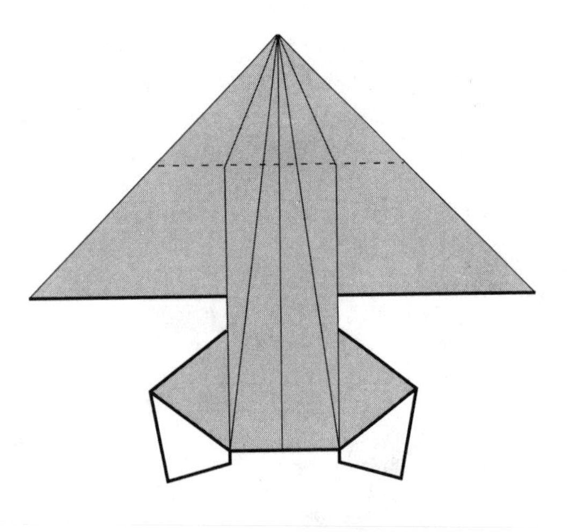

17. Fold down the tip of the plane so it lies at the same level as the bottom edge of the wings. Make certain the tip is in the exact center, and crease the front.

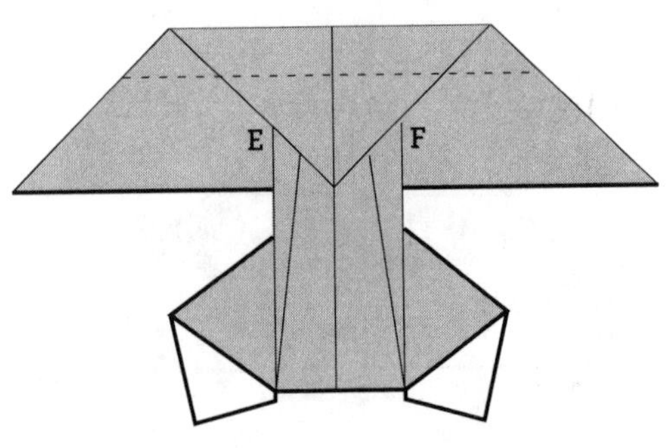

18. Fold down the front edge again, so that the edge passes through the points labeled E and F. Keep the central crease aligned. The result should look like Fig. 19.

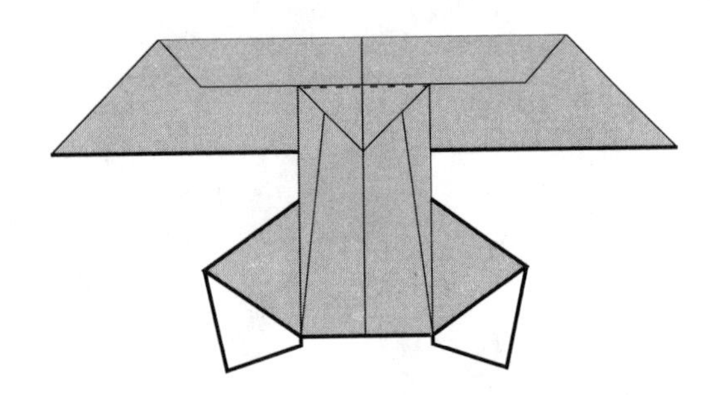

19. Fold the tip back up over the edge, as shown.

Turn the plane over, gently recreasing the central crease as you do.

20. Fold up a tail by making an accordion fold, in the area shown. Crease well. *See* the illustration of the finished plane for detail.

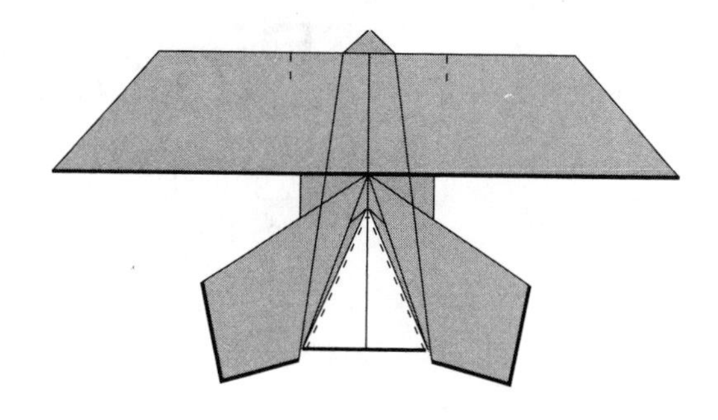

Gently bend each wing downward only on the forward edge, at the point shown. Recrease the front edges of both wings, making them as flat as possible.

Flying instructions for PLANE 9 are on page 41.

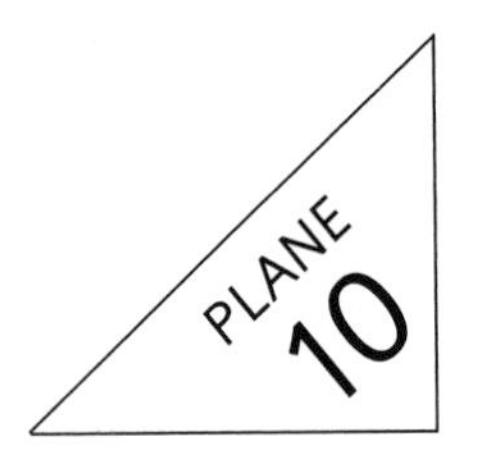

Biplane

A BIPLANE is a plane with two sets of wings, one directly above the other—these were common in the early days of aviation, and are still occasionally seen. It seems farfetched to try to fold one from a single sheet of paper, yet it can be done, as this design demonstrates. This plane employs a couple of complicated folds that make it a little difficult for the beginner, and a challenge even for the experienced. Follow the directions carefully, and you will see it take shape, and with a little luck, take to the air. (Folding instructions are on pages 51–56.)

To fly

Hold the plane from behind. Push forward gently, and release.

Adjustments

Make certain the wingtips are folded up at the same angle on each side, and approximately the same distance from the tip. If the plane veers to one side, try steering as directed in the introduction, or adjust the flaps under the wings.

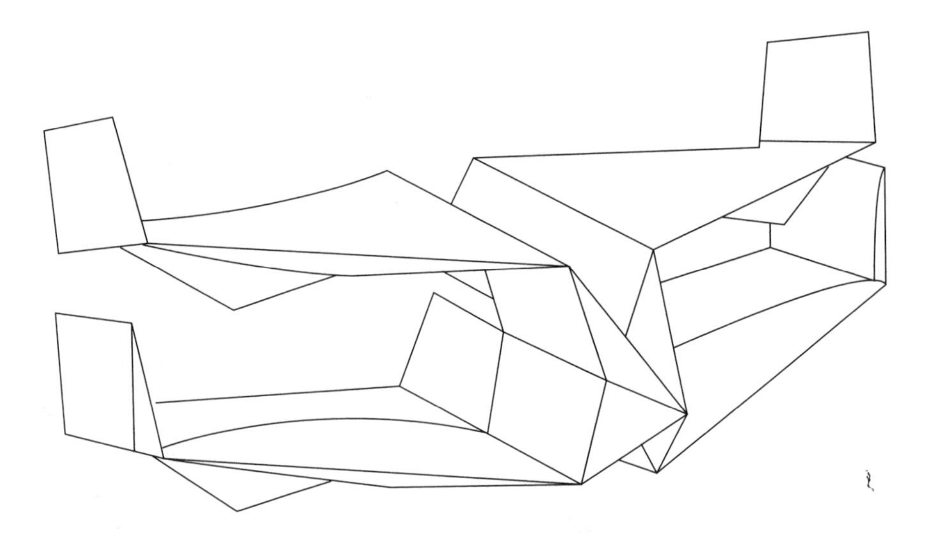

1. Make a central crease by folding one edge of the paper to the opposite edge. Crease very well.

Unfold.

Make a similar crease crosswise to the first, again by matching up opposite sides of the paper. Crease very well.

Unfold.

Turn the paper over.

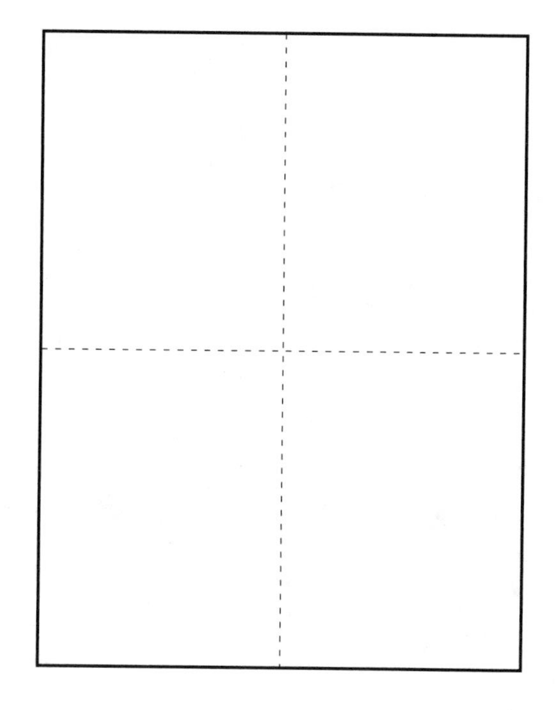

2. Make a diagonal fold as shown. This is done by folding over the point where the two creases intersect, then lining up point A exactly with the vertical fold at the bottom of the paper. It should look like Fig. 4 in reverse. Make certain your fold passes through the exact center of the paper where the creases intersect, and crease very well.

Unfold.

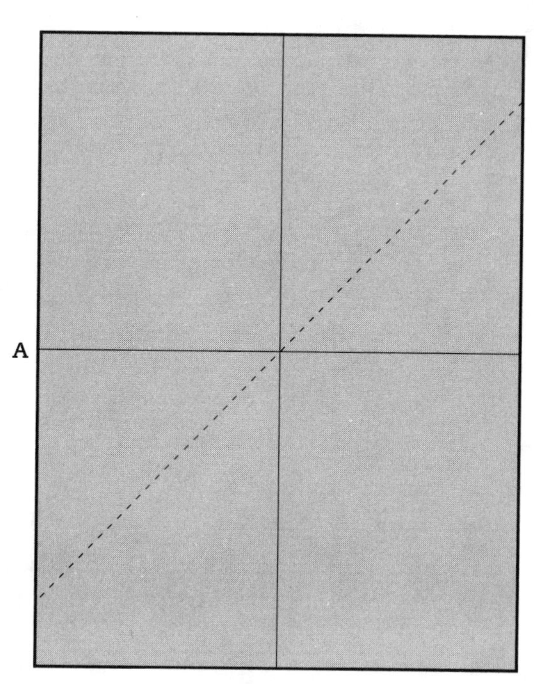

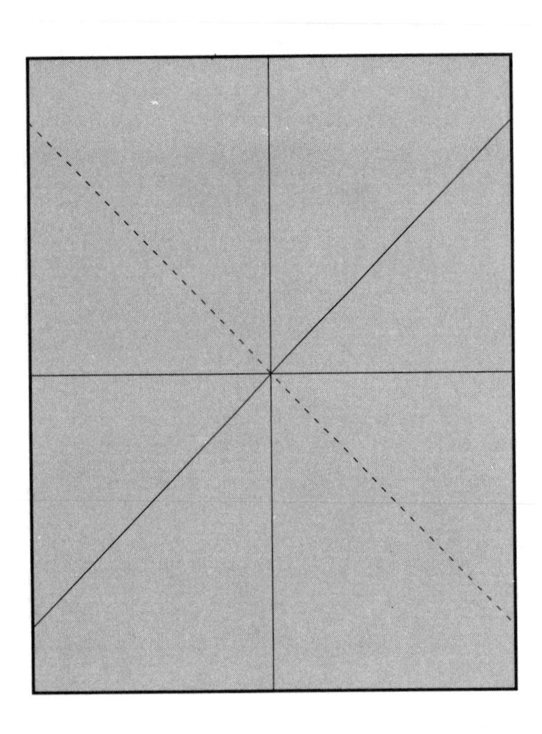

3. Make another similar diagonal crease, crossing the one you just made in the exact center. Crease well. Do not unfold.

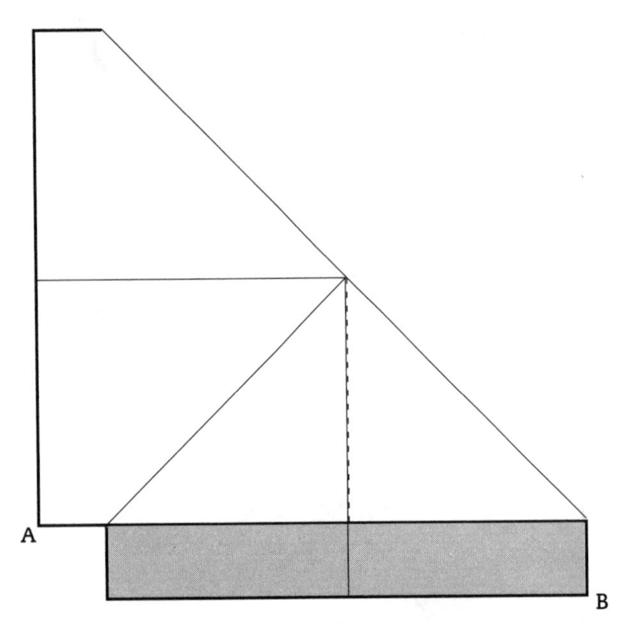

4. Using the creases already made, you will now make an accordion fold by lifting the corner labeled A and bringing it over to match the corner labeled B. The rest of the paper should follow, forming the shape shown in Fig. 5. Recrease all folded edges, making certain that the upper layer of paper exactly matches the lower layer.

5. Place the paper as shown. Fold the double bottom edge of paper forward so it is approximately ½ inch from the tip, taking care to keep the central crease exactly aligned. When you are sure the fold is exactly where you want it, run your thumbnail over the fold against a smooth, hard surface to make the crease very tight.
Turn the paper over, placing it as in Fig. 6.

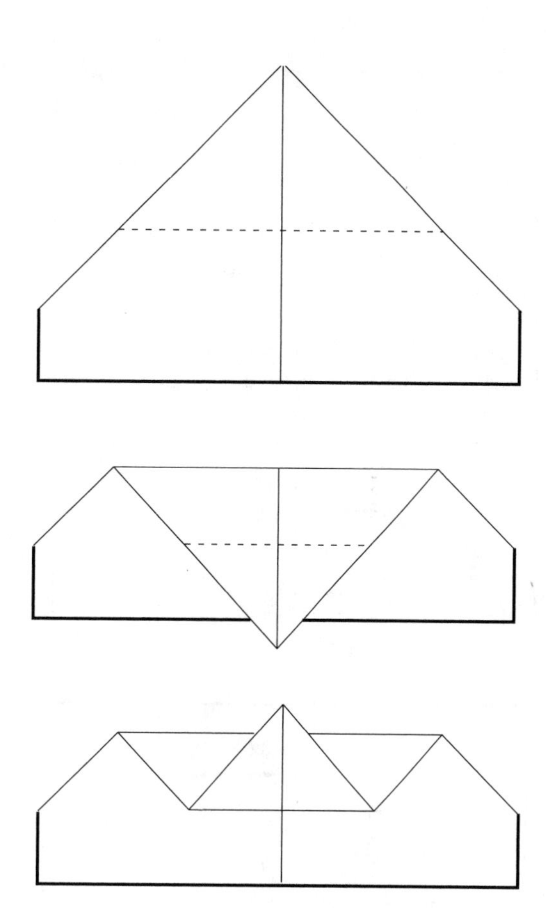

6. Fold the tip of the paper forward so it is approximately ¾ inch beyond the front fold, keeping the central crease exactly aligned. Reinforce this crease in the same way as the last, running your nail along the fold over a firm surface.

7. The result should look like this. Completely unfold the paper, so that the center is bulging out toward you.

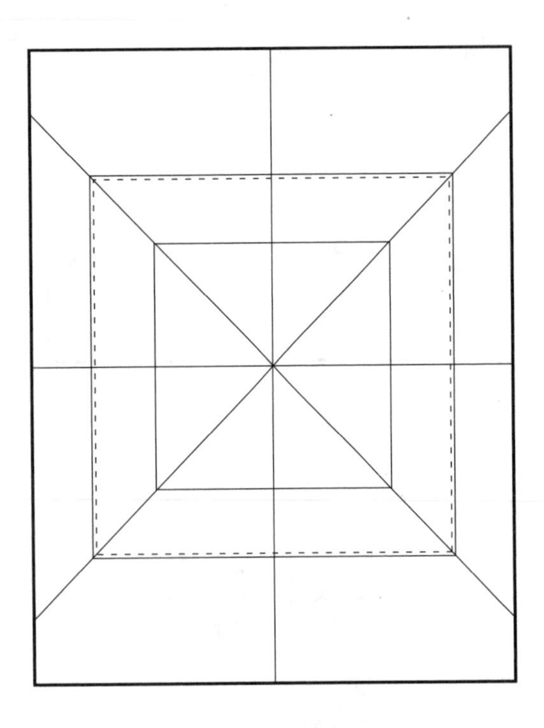

8. The paper should look like this, with lines intersecting in the center, surrounded by two boxes, one small and another large. You will use these boxes to form a telescoping fold, by turning the section between them inside-out. Start by creasing all around the larger box, so that it all bends in the same direction with its crease towards you.

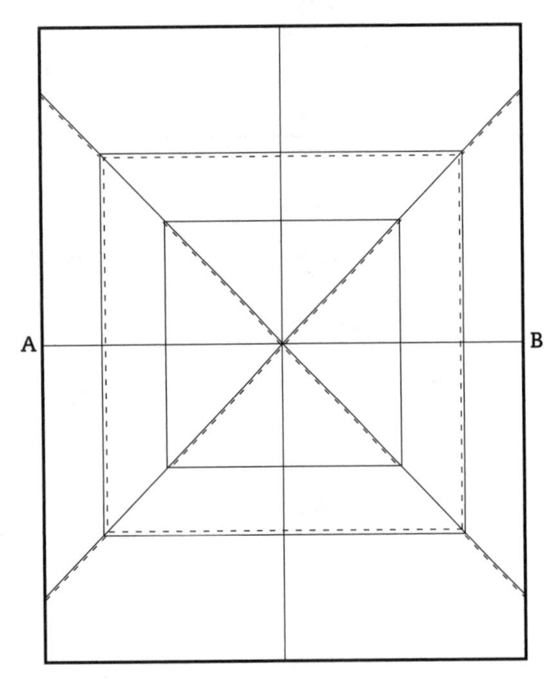

9. Now make the telescoping crease. Start by pinching each of the diagonal creases near the corners so they bend towards you outside the large square. Between the large and small squares, this fold should be reversed, and the small square should all bend in the opposite direction as the large one, so that the center of the paper remains pointing toward you. (In the drawing, creases which should bend toward you are shown with a broken line.) Push point A and B towards the center of the paper.

10. This drawing shows the telescoping fold in progress. Make certain the paper folds exactly along the creases, adjusting if the paper tries to bend in other places. When you are certain all the bends are along the creases, flatten the plane so it looks like Fig. 11, and crease well.

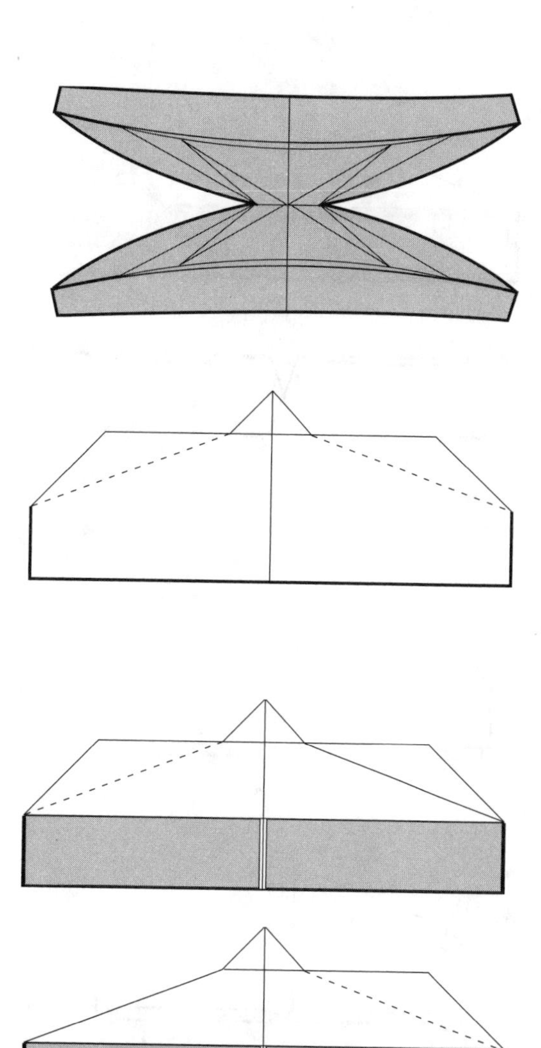

11. Note that the plane has two sets of wings, the top ones resting exactly over the bottom ones. Crease down the forward edge of the top wings only, in the manner shown. Crease very well.

Take the left top wing and fold it over the right, exposing the left bottom wing, as seen in Fig. 12.

12. Make a similar fold as shown on the left bottom wing, creasing very well.

Fold both top wings over to the left, exposing the right bottom wing.

13. Fold down the front of the right bottom wing in exactly the same way, creasing very well.

Now, keeping both top wings on the left, lift the plane and fold the left bottom wing back to the right, so that both wings formerly on the bottom are on the right, and both wings formerly on the top are to the left, as shown in Fig. 14.

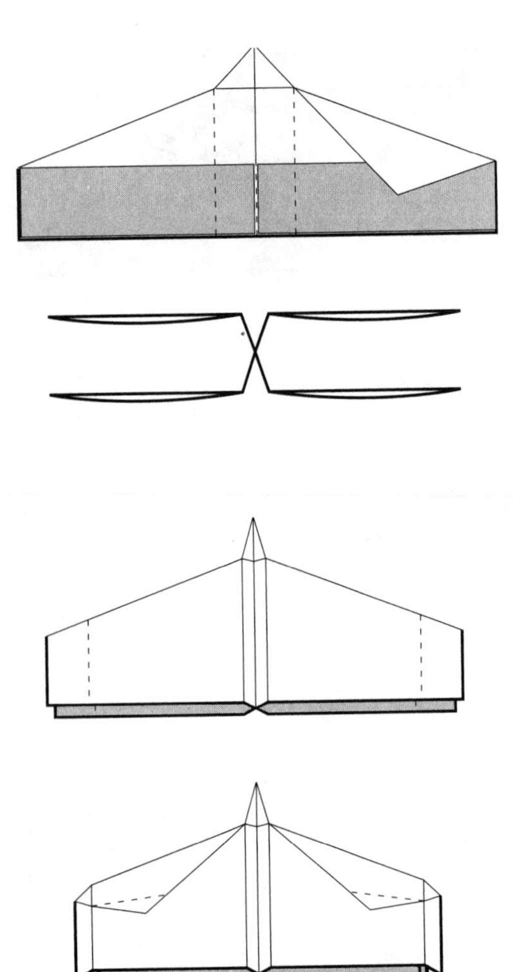

14. Fold each of the upper wings toward you, as shown. The one shown on the right will become the top wing, the one shown on the left will become the bottom. Turn the plane over, and repeat the fold with the two wings on the opposite side.

15. Adjust the wing folds so the plane looks like this from the front—the upper and lower wings parallel to each other, with the wing folds facing downward.

16. This is a top view of the plane. Recrease the front edges of all four wings between your fingernails. Fold up the ends of each wing, both upper and lower, about ¾ inch from the wingtip, and parallel with the body of the plane.

17. This is a bottom view. On each wing, fold a small flap away from the wing, as shown. This must be done on the upper and the lower wings. Check the plane carefully to see that the wings on one side match those on the other side as exactly as possible.

Flying instructions for PLANE 10 are on page 50.

PART 2

Jets

SLEEK and sweptwinged, the jet aircraft was a relative latecomer in the history of aviation; yet for those of us who grew up in the early days of the jet era, it was the sweptback wings of the supersonic jet, still fresh in the air, that our first dart-like paper airplanes most resembled, and as we flew them, it was the lion's roar of the jet plane streaking through the sky that fired our imaginations. The folded paper airplane continues to lend itself very well to the jet theme, and the planes in part 2 are a sampling of the wide variety of effects that can be achieved. Paper jets, like their real-life counterparts, fly much faster as a rule than the straight-winged designs of part 1, and hence can be given a hard throw without difficulty; most can be flown out of doors, so long as the wind is not too strong.

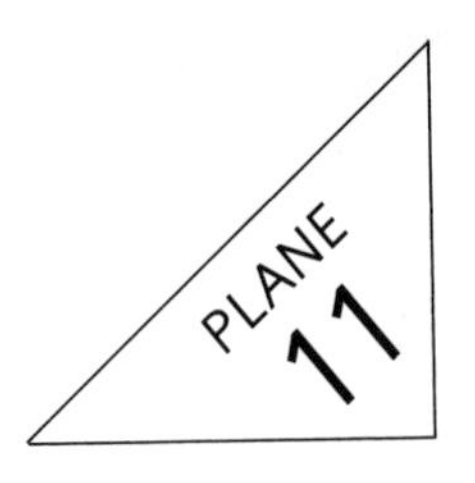

Hornet

THIS DESIGN is adapted from the traditional dart-like jetplane that was once the mainstay of paper airplane folding. It is aptly named for the stinger configuration in the rear; an alternate design, more traditional and lacking the stinger, is also presented. (Folding instructions are on pages 59–62.)

To fly

Hold from beneath, and toss gently straight forward. This will enable you to make necessary adjustments. Once adjusted, the plane can be thrown forcefully when out of doors. Throw either forward or upward, as hard as you can.

Adjustments

The plane can be steered by bending the wingtips up or down. Some elevator might be required, especially on the alternate design. If, after a time, the plane performs poorly, it can be revived by folding the wings down against the body and recreasing all the edges.

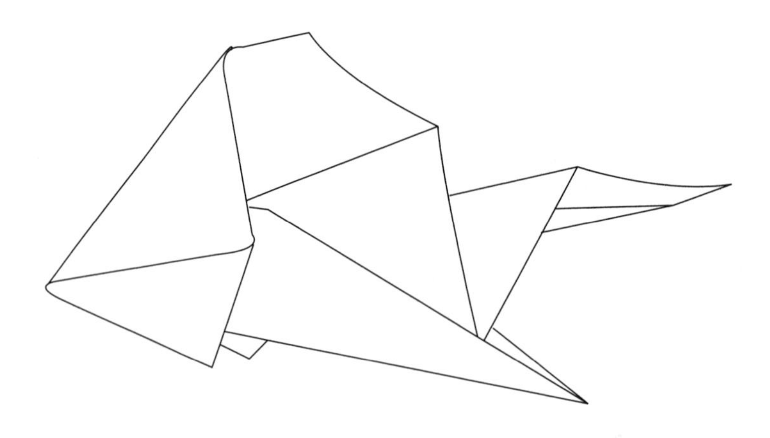

1. Make a vertical central crease, as shown, by aligning the left edge of the paper with the right. Crease well.

Unfold.

Turn the paper over.

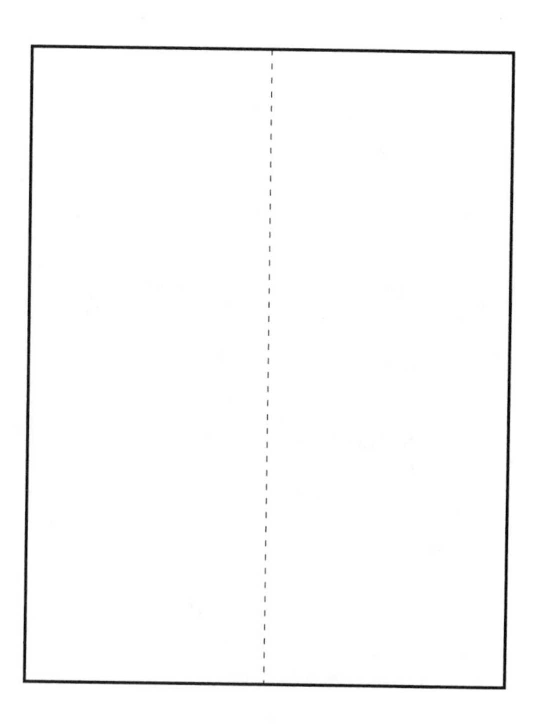

2. Fold down both top corners as shown, aligning the edges carefully against the central crease, as in Fig. 3.

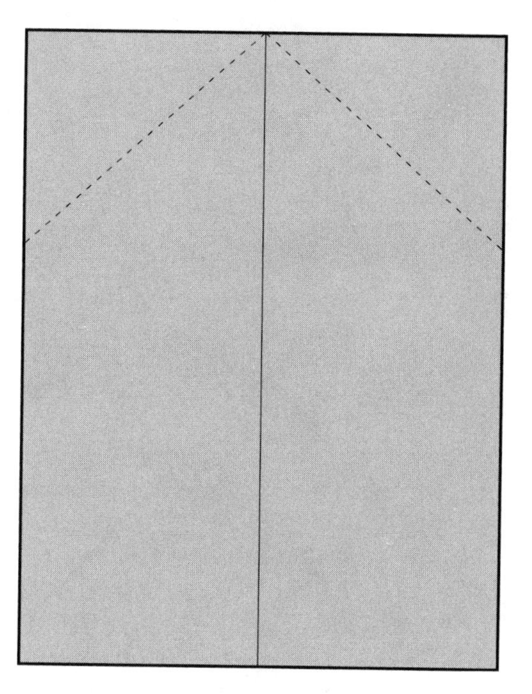

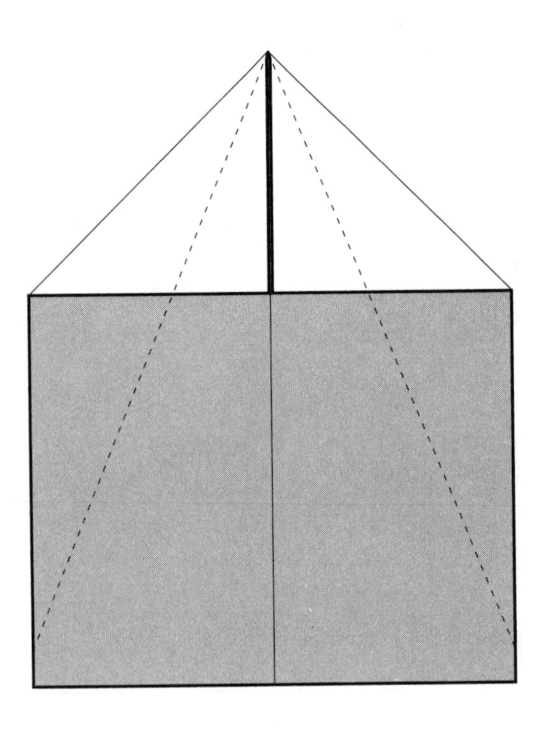

3. Make another similar fold, again aligning the edges on each side against the central crease.

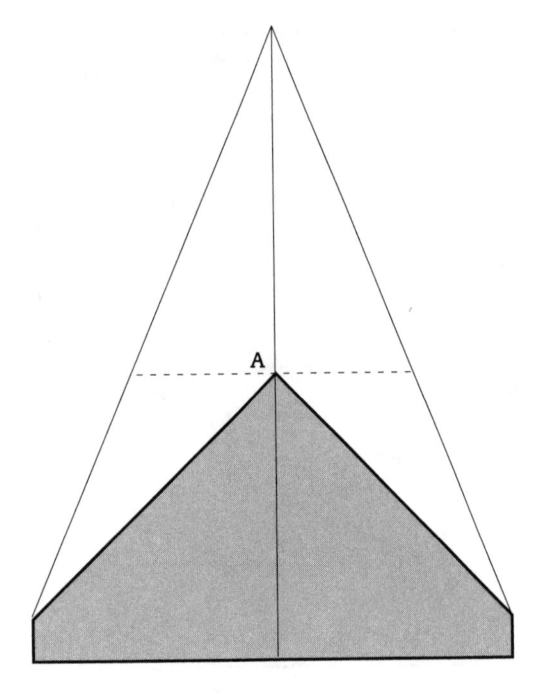

4. Shows the result of Step 3. Now fold down the tip, your fold passing through the point labeled A where the paper edges come together. Make certain the central crease is lined up exactly on the backside of the plane. Crease well.

5. Make diagonal folds on each side, folding the top folded edge to the central crease. The fold will follow the diagonal edges of paper on each side, as shown. For an alternate finish to the plane, go to Step 6(A).

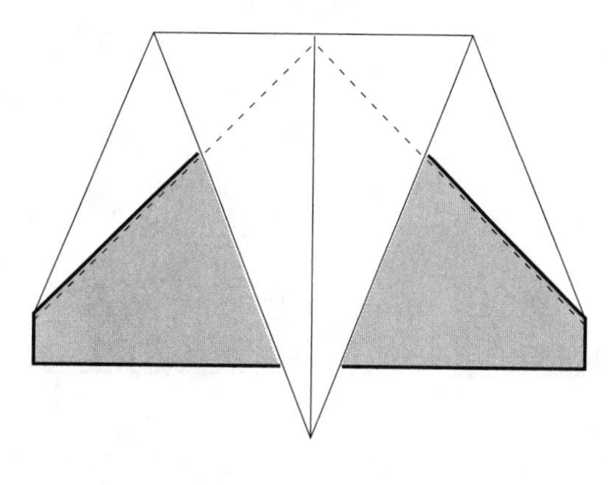

6. Shows the result of Step 5. Turn the paper over, and fold in half along the central crease.

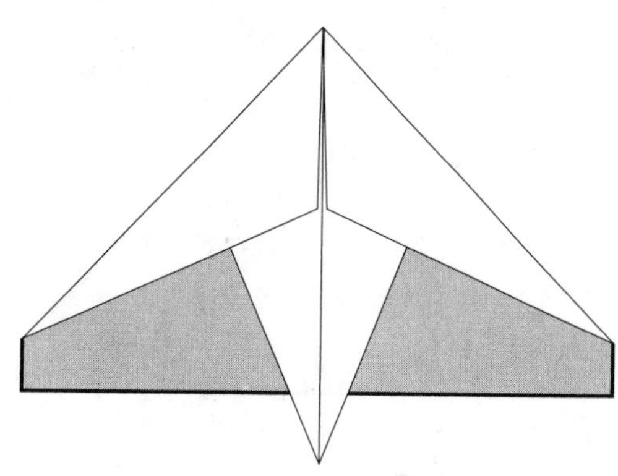

7. Fold the wing down as shown, aligning the edge of the wing with the central fold, from the tip back. Repeat with the other wing, aligning the wings exactly. Partially unfold. (*See* drawing of finished plane.)

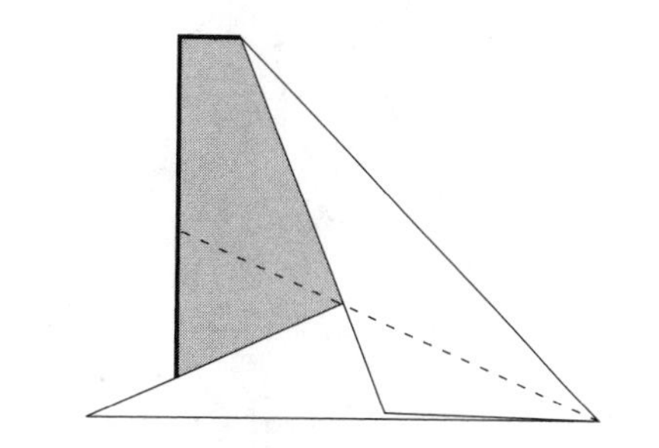

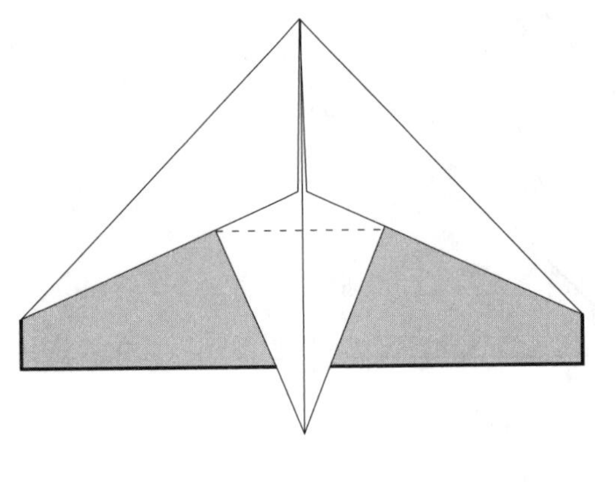

6(A). For an alternate finish without the stinger, after Step 5, fold the pointed end of paper forward as shown, so it touches the front tip of the plane exactly.

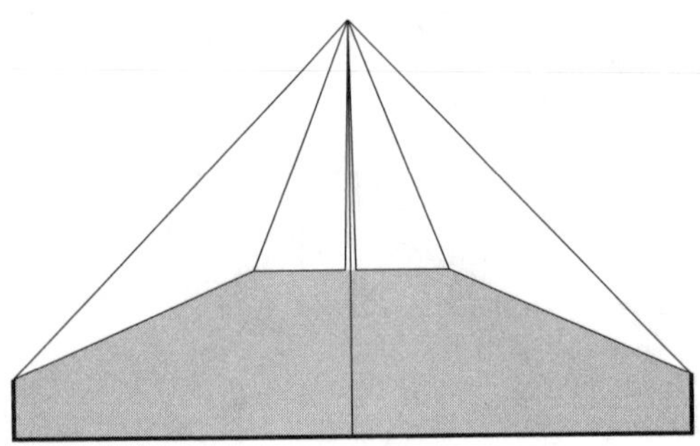

7(A). Shows the result of Step 6(A). Turn the paper over, then fold in half along the central crease. Make certain the wings align exactly with each other.

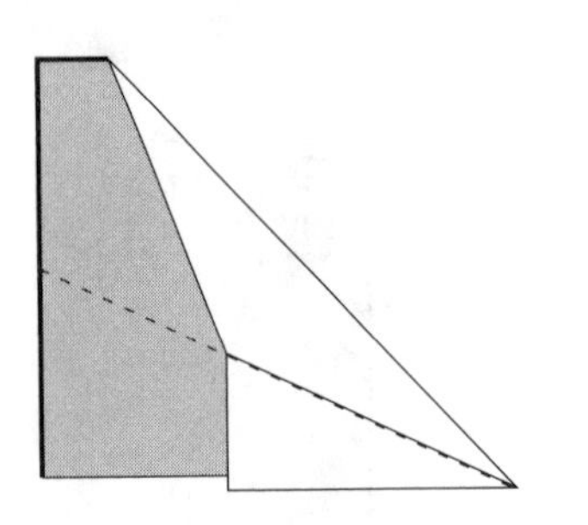

8(A). Fold down each wing, following the edge of the pointed section you folded in Step 6(A), as shown. Make certain both wings match exactly. Partially unfold, so the wings are extended equally.

Flying instructions for PLANE 11 are on page 58.

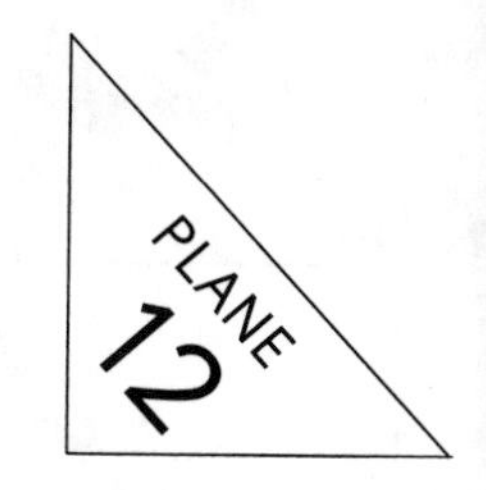

Aerobatic jet

THIS PLANE is rather attractive with its almost rounded, swept-back wings, and is folded in an unusual way. With a good skyward toss it will make long, soaring outdoor flights. (Folding instructions are on pages 64–67.)

To fly

Hold the body of the plane from below, and throw gently forward. When well adjusted, try throwing as hard as you can, either straight forward or upward; you can even try straight up. With practice, the plane can be made to climb, level off, and make a long, slow descent.

Adjustments

See that the wings are equally folded. If the plane tends to dive into the ground, bend the back edge of the wings up slightly. If it tends to turn to one side, the back edge of the two wing fins can be bent as rudders to correct the turn (careful—a little bend goes a long way). If the plane does not fly at all or goes into a tailspin, check the symmetry of the plane as described in the subsection entitled "Making it fly" in the introduction, and make certain all folds are symmetrical.

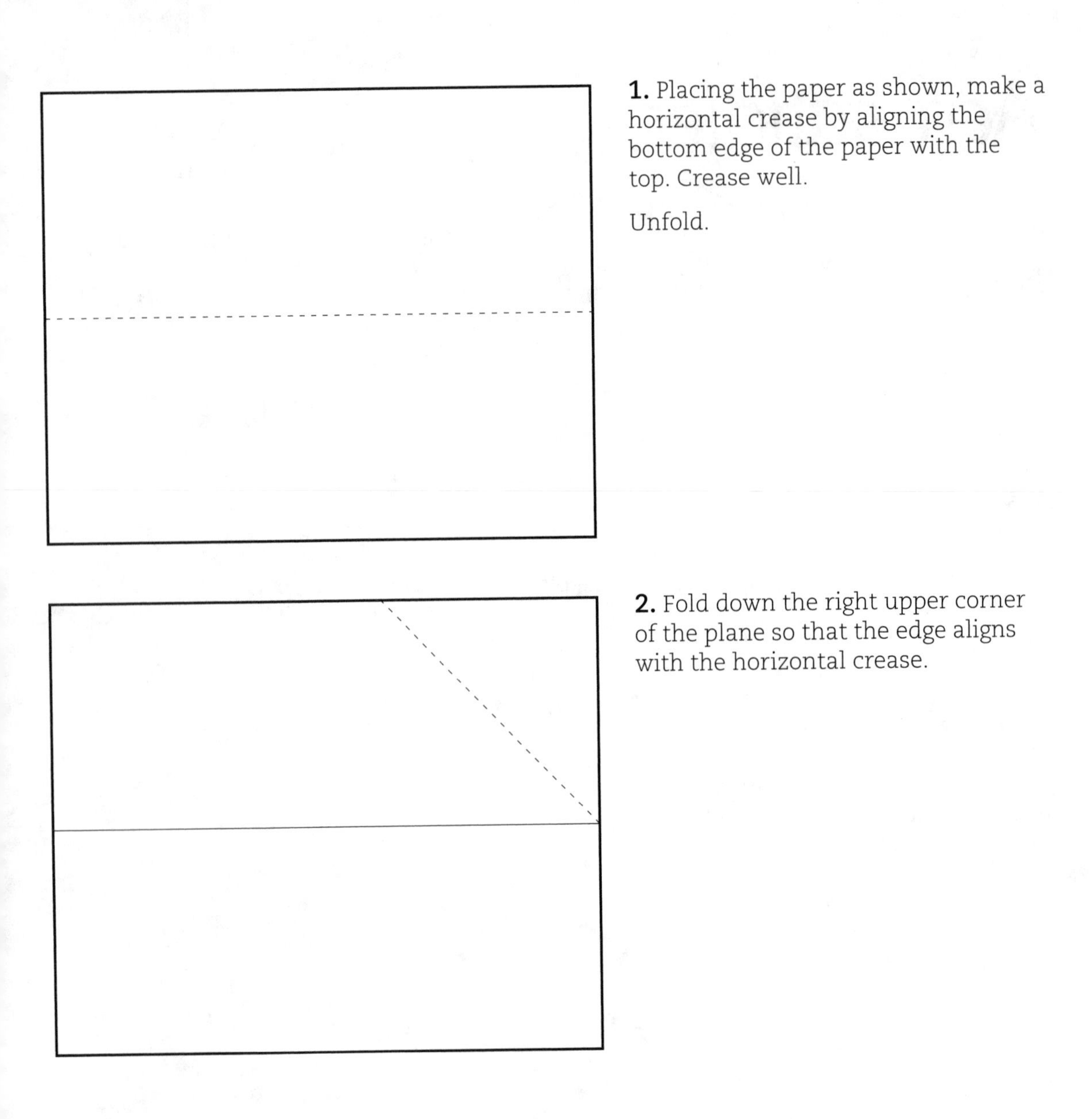

1. Placing the paper as shown, make a horizontal crease by aligning the bottom edge of the paper with the top. Crease well.

Unfold.

2. Fold down the right upper corner of the plane so that the edge aligns with the horizontal crease.

3. Make another similar fold, aligning the diagonal edge with the horizontal crease.

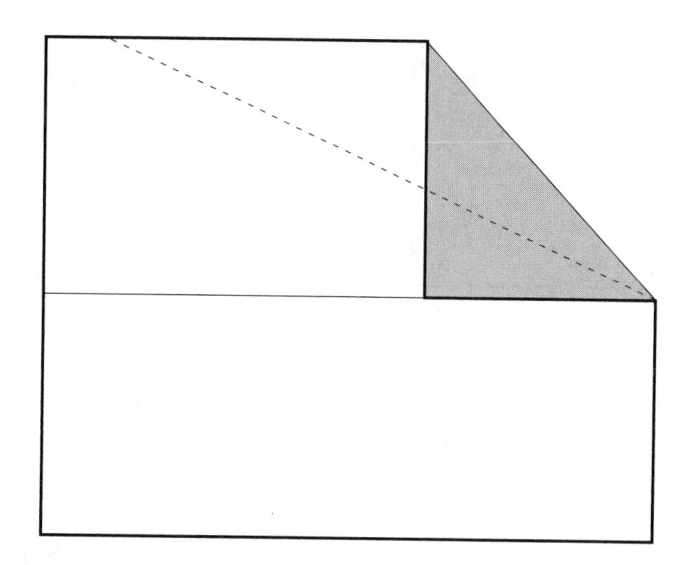

4. Shows the result of Step 3. Now fold up the bottom half of the paper along the horizontal crease. Crease well.

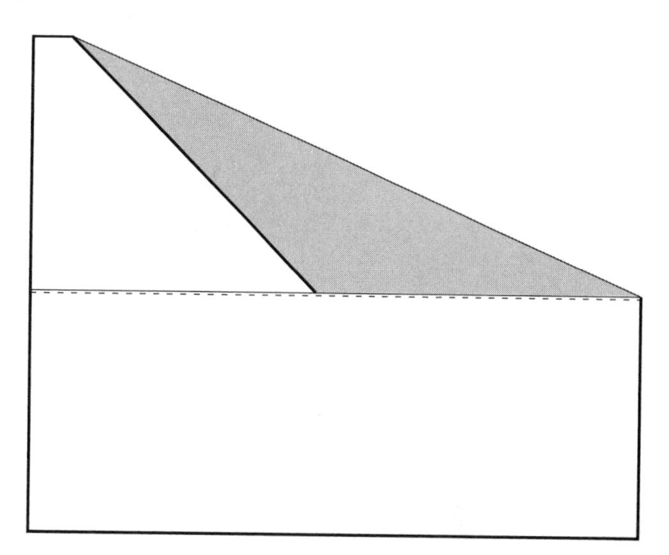

5. Fold down the left upper corner, aligning the edge with the bottom folded edge.

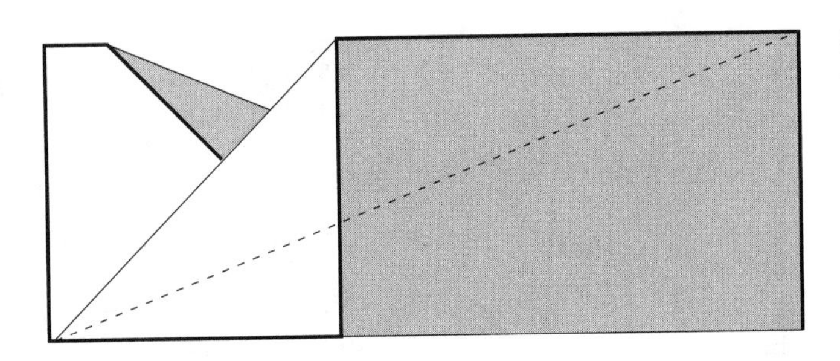

6. Fold down again, aligning always with the bottom folded edge.

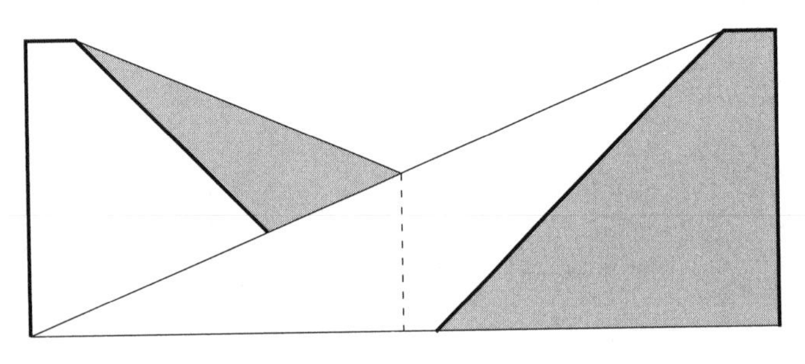

7. Shows the result of Step 6. Now make a central crease as shown, by folding over the right half of the plane on the left half. Make certain both sides match perfectly, and crease well.

Unfold, then rotate the paper to the position shown in Fig. 8.

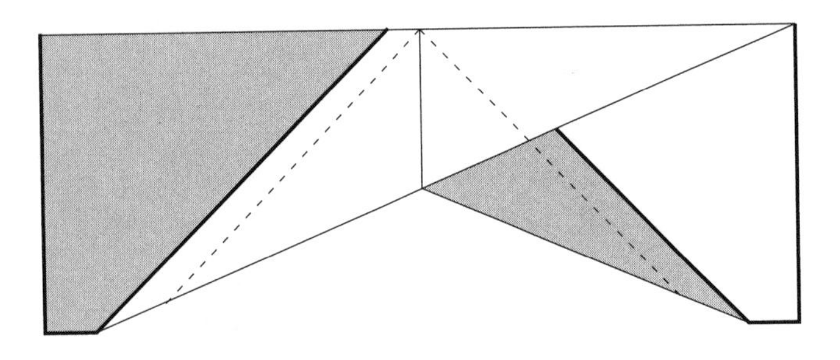

8. Fold down first one wing, then the other as shown, aligning the top edge with the central crease. The result should look like Fig. 9.

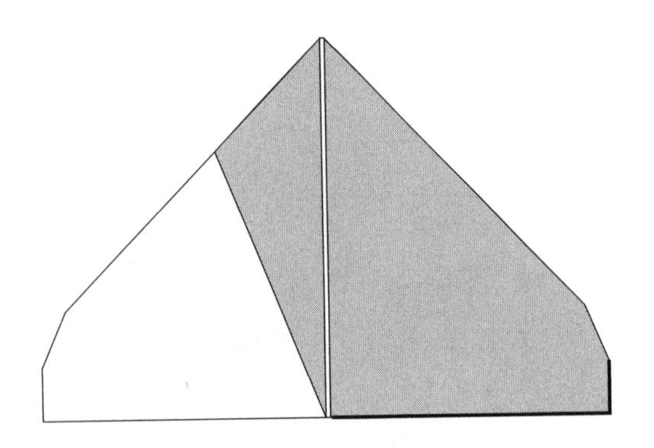

9. Shows the result of Step 8. Fold in half along the central crease.

10. Fold down each wing, approximately ¾ inch from the central fold, and parallel to it. Be certain you fold both wings in exactly the same spot.

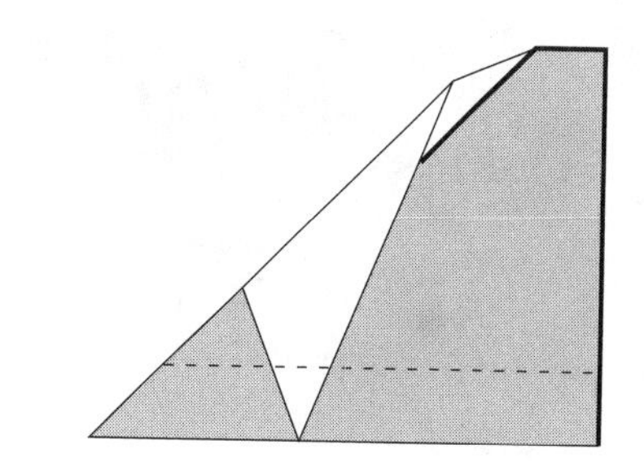

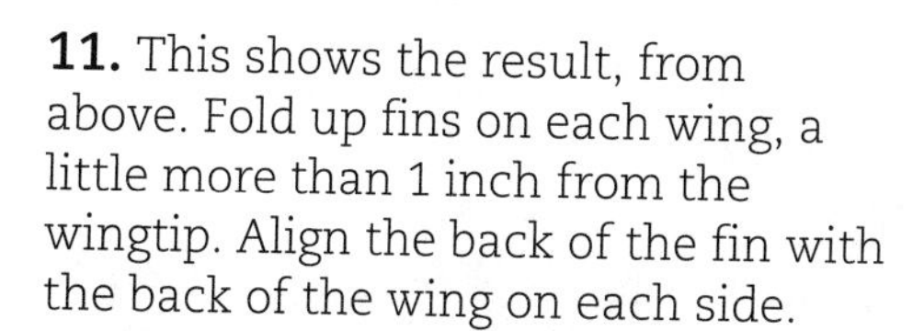

11. This shows the result, from above. Fold up fins on each wing, a little more than 1 inch from the wingtip. Align the back of the fin with the back of the wing on each side.

Fold up both central corners as shown, to form a small tail. Partially unfold.

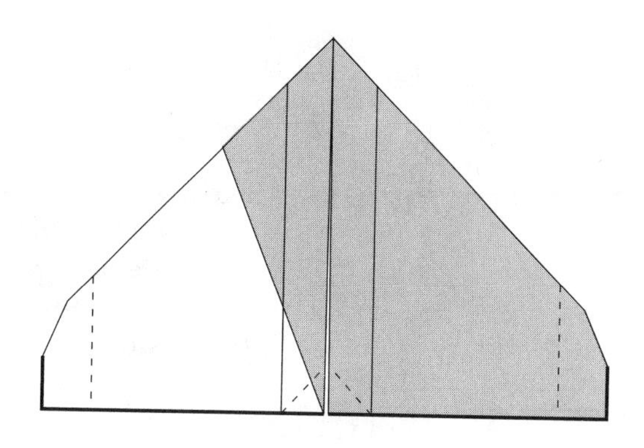

Flying instructions for PLANE 12 are on page 63.

Twin fin

THIS UNUSUAL jet plane is folded in a rather unconventional way and gains its name from the two vertical flaps that jut upward from the base of the wings. The plane flies well indoors and out. (Folding instructions are on pages 69–74.)

To fly

Hold the body of the plane from beneath, and toss firmly forward.

Adjustments

Make certain that the wings are slanted slightly upward from the body when the plane is in flight. If the plane tends to veer to one side or the other, use the back edge of the wings to correct the flight as directed in the subsection of the introduction entitled "Making it fly."

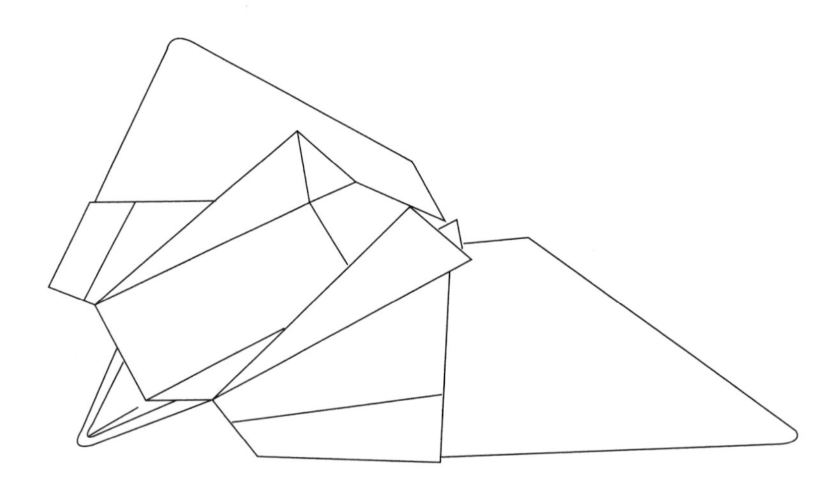

1. Make a vertical midline crease as shown by aligning the left edge of the paper exactly with the right. Crease well.

Unfold.

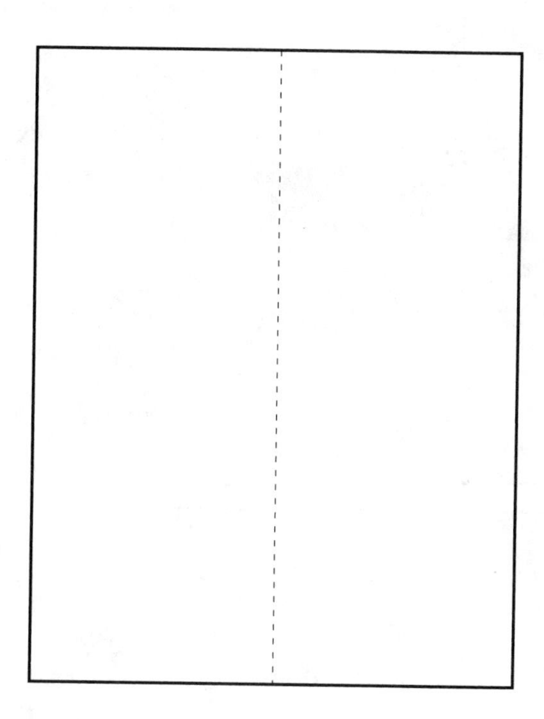

2. Make a diagonal fold as shown, by placing the left upper corner exactly on point A, at the center of the bottom edge, as shown in Fig. 3. Do not unfold.

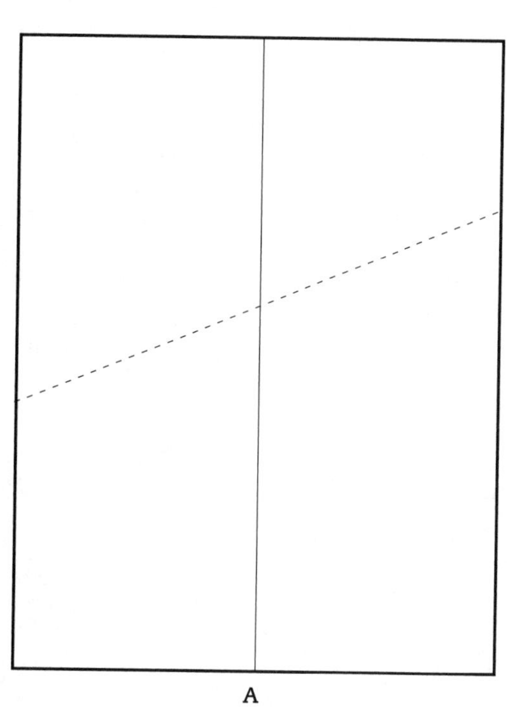

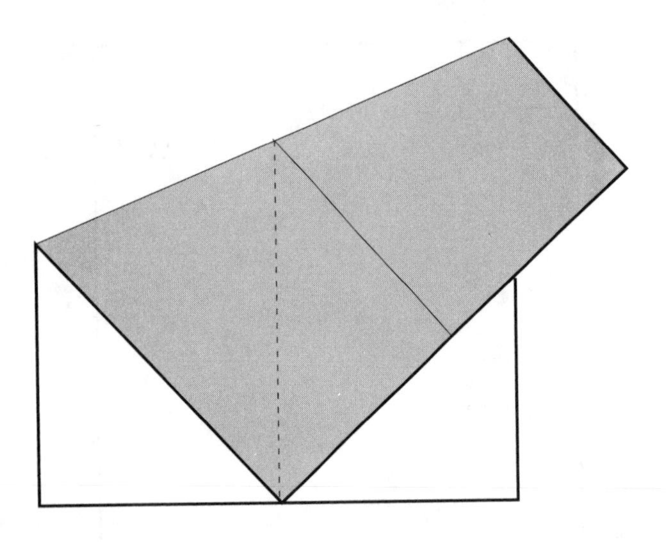

3. Carefully fold the left side over along the vertical crease, aligning the left lower corner with the right. Crease very well.

Unfold the previous two folds, opening the paper out completely.

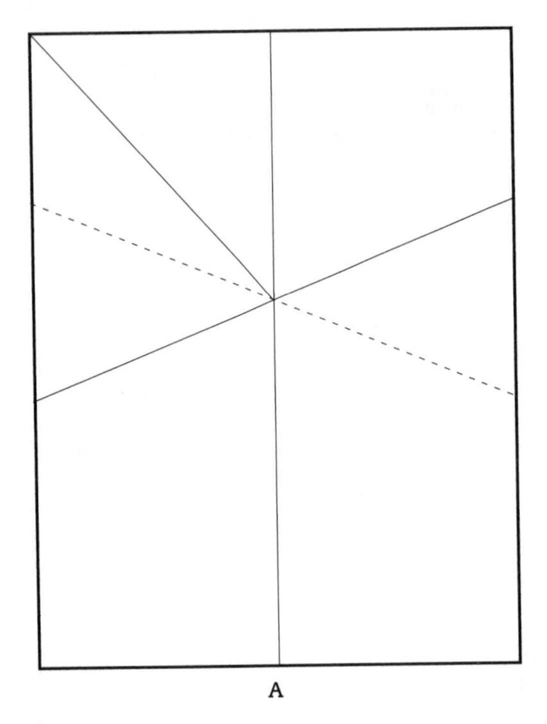

4. Now fold diagonally again as shown, bringing the right upper corner exactly to point A, as shown in Fig. 5.

5. Carefully fold the right side of the paper over, aligning the right lower corner with the left. Crease well. Unfold only the last fold.

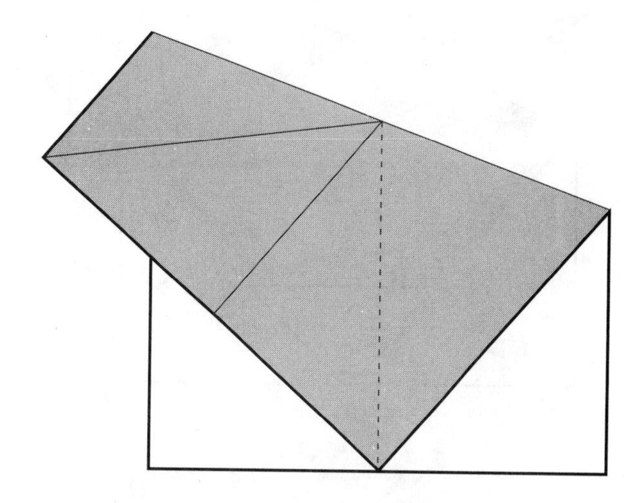

6. Holding the right side of the paper flat with your hand, grasp the corner marked point A and slide point A toward point B until it is next to point B. The paper will naturally fold outward along the creases shown, causing point C to jut out toward you, as in Fig. 7.

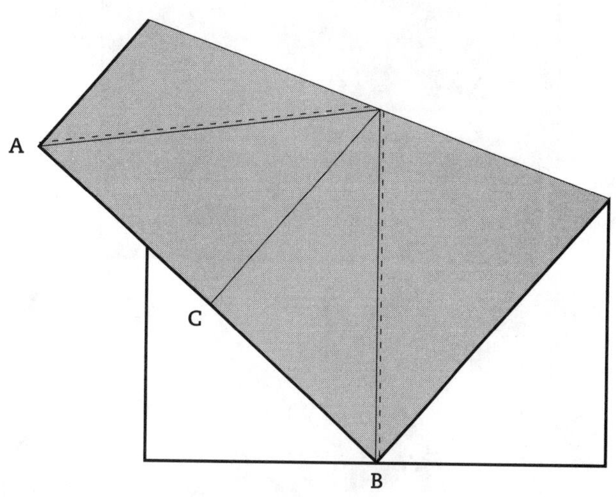

7. Note that there is a triangular section of paper sticking up vertically from the flat paper, with point C at its tip. Carefully press this section flat with your fingers, working from the front back and keeping point C exactly in line with the central crease. When you have the section well-placed, crease the edges of the section from front to back and from middle to outside (*see* Fig. 8).

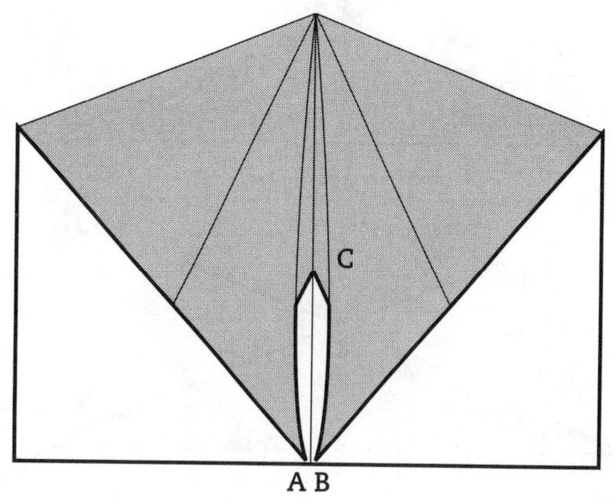

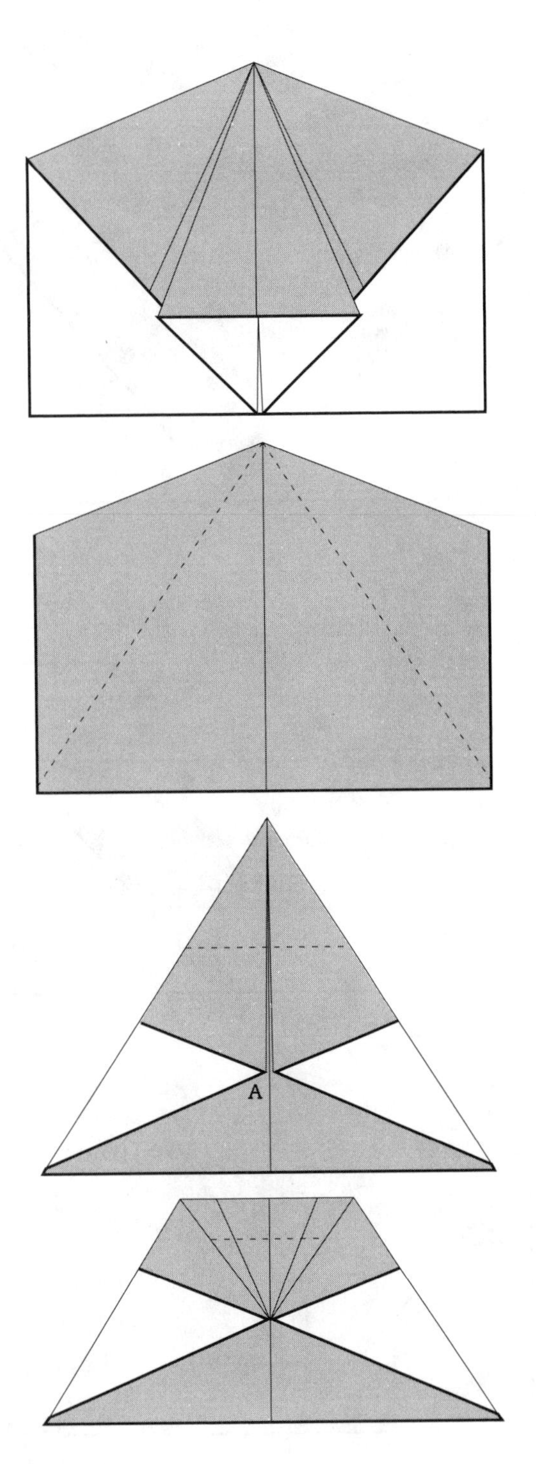

8. Shows the result of Step 7. Turn the paper over.

9. Make a fold on each side as shown, so that the edges line up along the central crease as in Fig. 10.

10. Fold the tip down to point A, keeping it exactly aligned with the central crease. Crease very well because this edge will form the front of the airplane.

11. Fold the tip forward again so that the tip protrudes approximately ¾ inch beyond the front of the plane. Make certain the central crease is aligned, and crease.

12. Shows the result of Step 11. Turn the paper over.

Fold in half along the central crease, aligning the wings carefully with each other.

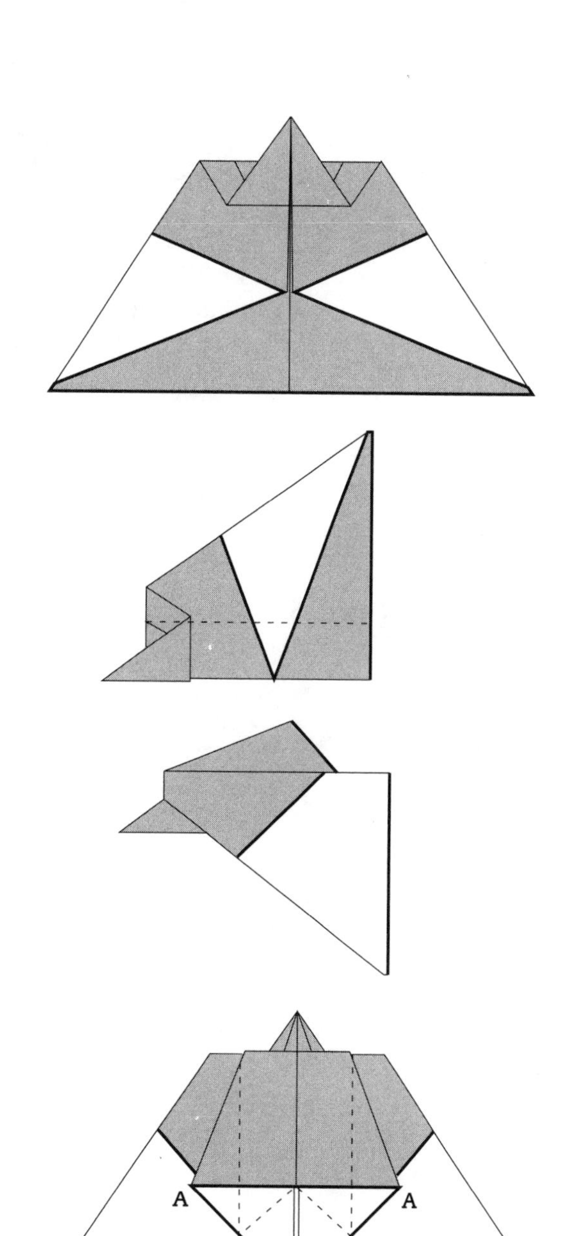

13. Fold down each wing as shown, approximately 1 inch from the central crease and parallel to it, as in Fig. 14.

14. Note that the two fins remain standing upward, and that the wing fold extends from the front of these fins. Match both wings exactly, and crease well. Open the plane out, positioning it as in Fig. 15.

15. Fold up the backward-pointing tips on each side as shown, matching the tip with point A on each side. Unfold.

Fold up each fin along the wing fold, as shown.

Flying instructions for PLANE 13 are on page 68.

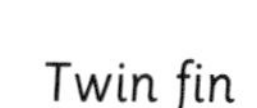

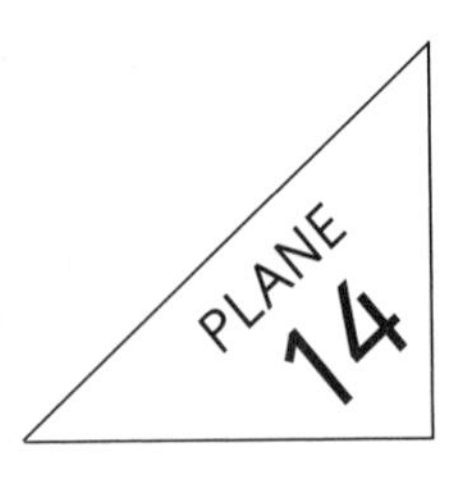

Sleek jet #1

THIS ELEGANT DESIGN looks as if it ought to fly at 1,000 mph; in fact, the flight is smooth and relatively slow, and generally must be confined indoors. (Folding instructions are on pages 75–80.)

To fly

Hold from beneath to throw. Make a firm, level toss. This plane will fly long distances indoors, but will not tolerate outdoor breezes very well.

Adjustments

The main problem with this design is the tendency to stall, which is easily corrected with the wing bend described in Step 10. The same bends can be used to steer the plane if it tends to deviate to one side.

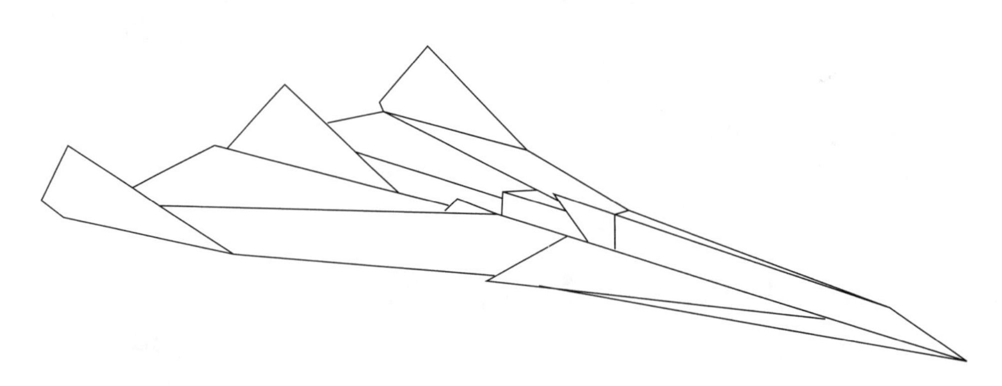

1. With the paper as shown, make a vertical midline crease by aligning one edge of the paper with the edge opposite.

Unfold.

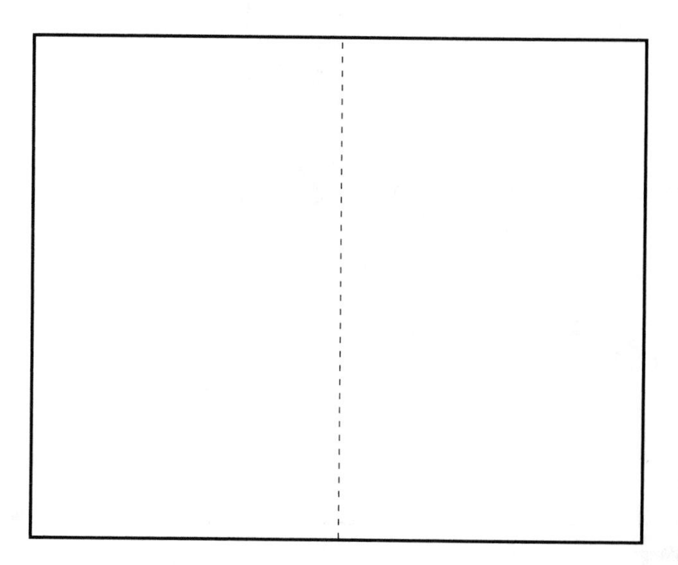

2. Fold down one top corner, aligning the top edge with the central crease. Repeat with the other top corner. The result should look like Fig. 3.

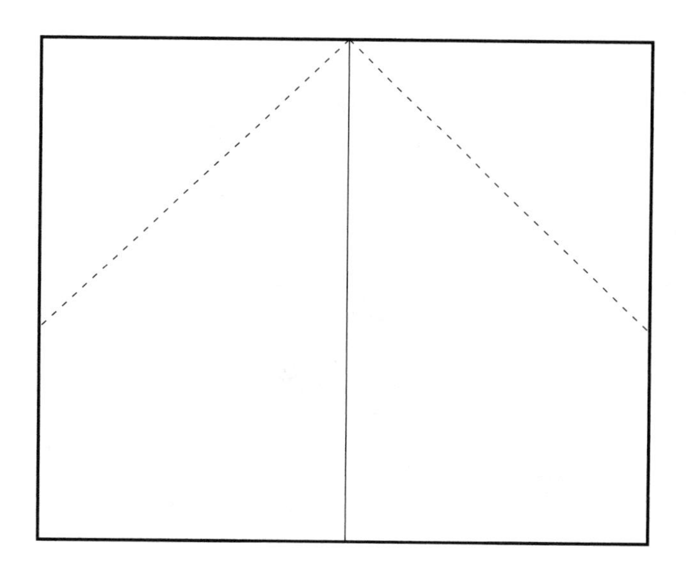

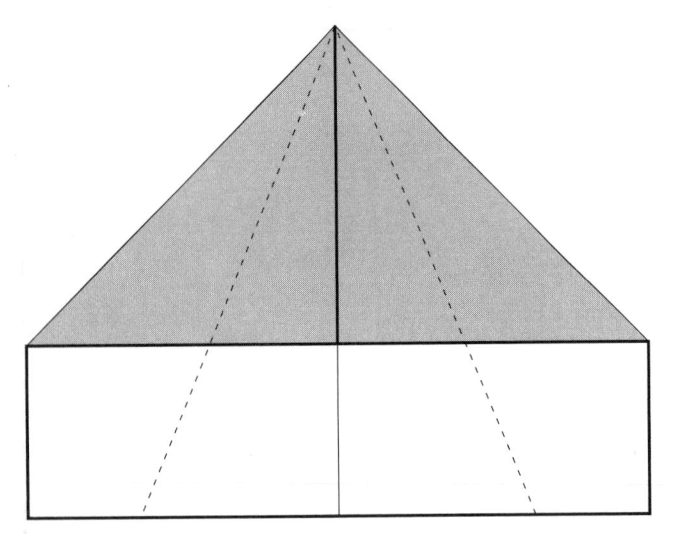

3. Make a similar fold on each side, again aligning carefully with the central crease. *See* Fig. 4.

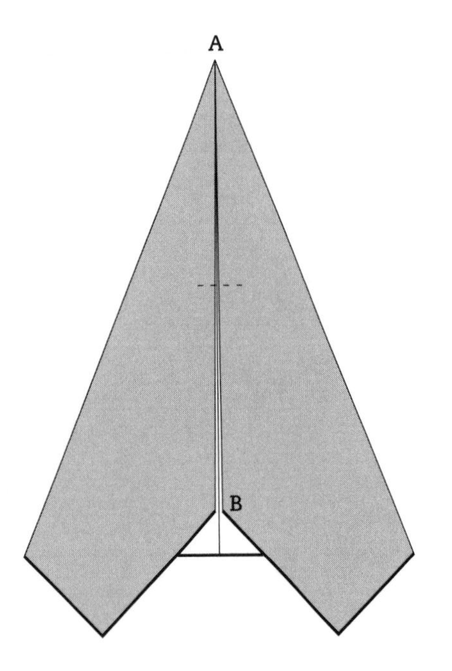

4. Shows the result of the previous fold. Now make a small crease by bringing the tip of the plane, labeled A, down to point B. Crease slightly across the central crease with the tip of your finger, just so the crease is visible. (Do NOT crease all the way across the plane.)

Unfold.

5. Now, fold a wing outward on each side. The fold should extend from the small crease at the top to the lower corner at the bottom, as shown.

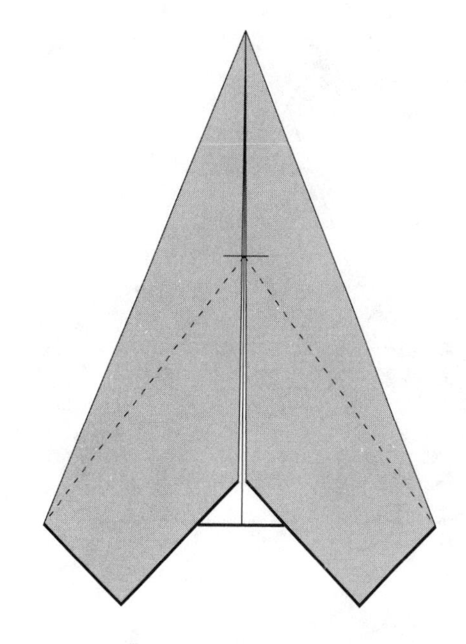

6. Shows the result of Step 5. Turn the plane over.

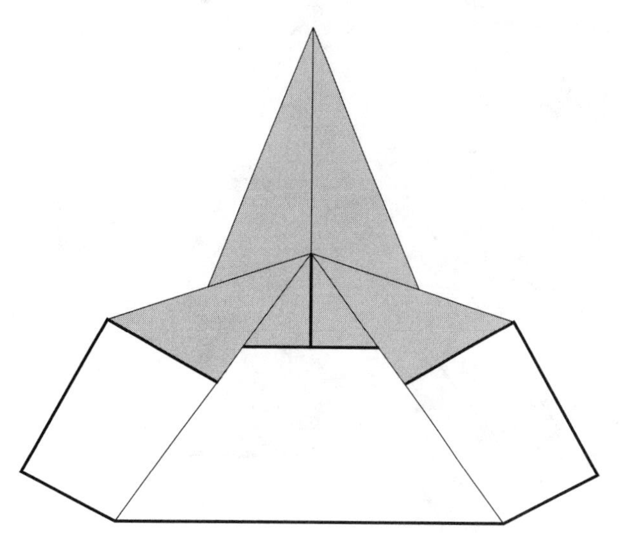

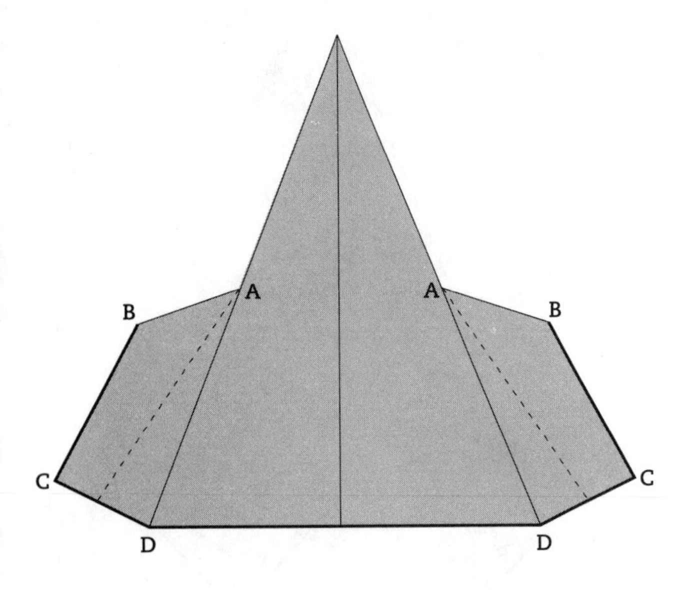

7. Now fold down each wing edge as shown. Start the fold at point A, folding so that edge BC crosses over point D. Make certain that both sides are folded exactly the same. Crease very well. The result should look like Fig. 8.

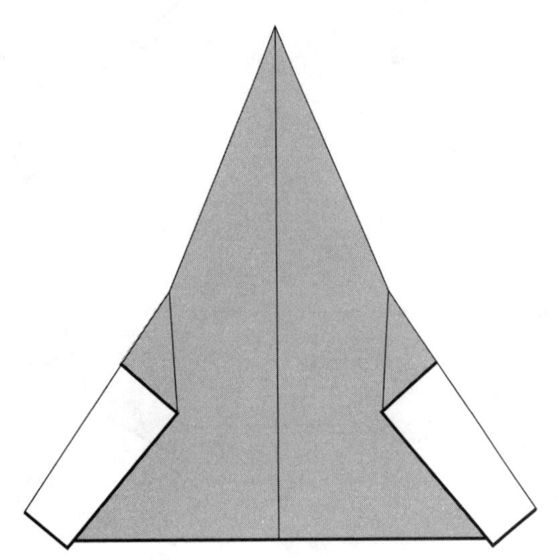

8. Shows the result of Step 7. Turn the plane over.

Fold in half along the central crease.

9. Fold down each wing as shown, starting approximately 1 inch from the tip and tapering to approximately 1 inch from the central fold at the tail. Make certain you fold both wings exactly the same.

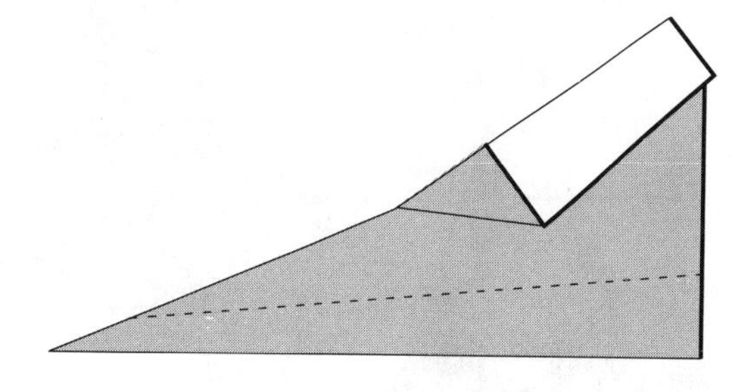

10. Make an accordion fold as shown by pushing the tail upward between the wings, then creasing from the sides.

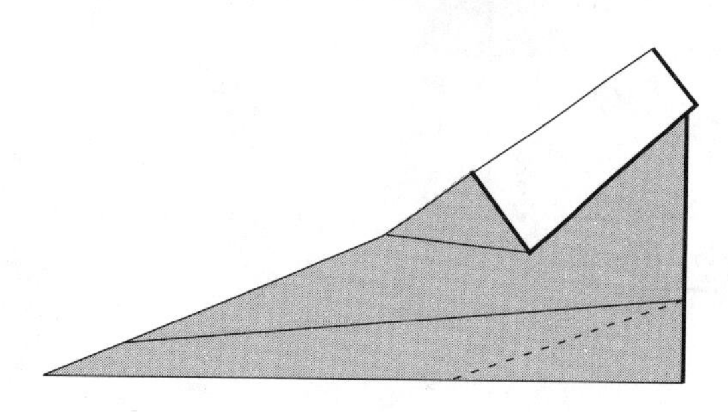

11. This is a top view of the plane showing the tail fold. Fold up the tips of the wings as shown, making the fold approximately parallel to the wing fold.

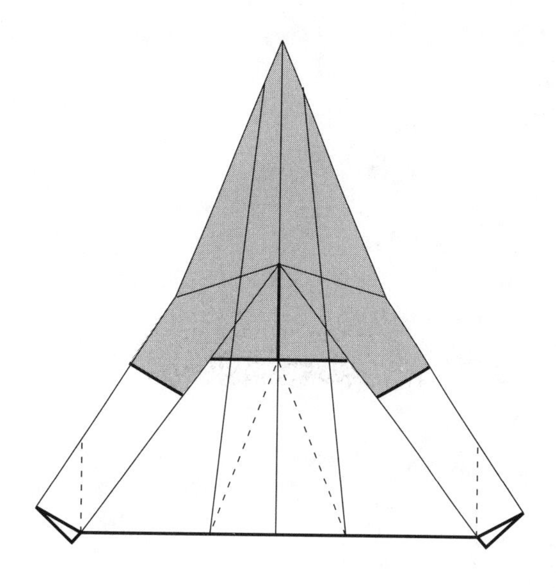

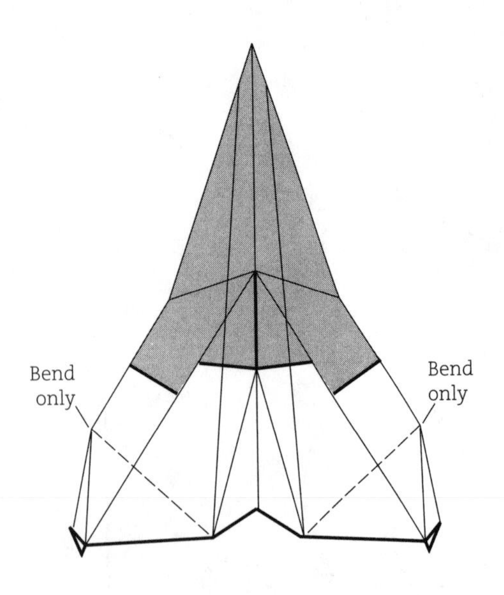

12. To avoid the tendency to stall, bend down both wingtips very slightly in approximately the place shown, in front of the wing fins.

Flying instructions for PLANE 14 are on page 74.

Sleek jet #2

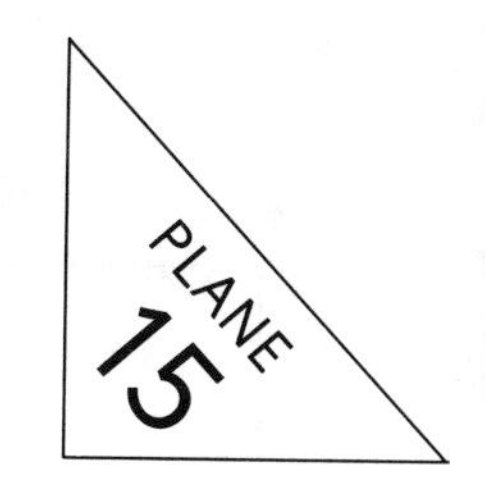

A SMALLER AND FASTER version of the previous design, this plane is intended for outdoor flight. It is a little more difficult to fold and might require some adjustment, but when you have it flying well, you will find it a true high-performance jet. The stronger your arm, the better the flight. (Folding instructions are on pages 82–86.)

To fly

Hold from beneath, and make a firm horizontal toss, adjusting as needed. When the plane is adjusted, take out of doors and throw as hard as you can, either straight forward or upward. If there is a breeze, throw into the wind. Adjust as you go.

Adjustments

If on your test flights the plane tends to dive, increase the upward bend from Step 14. You can increase this bend to make the plane fly in a loop and land. If the plane veers to one side or flips over, sight along the center of the plane from behind to see if both halves are exactly matched, adjusting as needed. It sometimes helps to fold the wings against each other and recrease all edges. (You can sometimes correct a turn by using the back end of the tail as a rudder, or by changing the angle of one or both of the wing fins.)

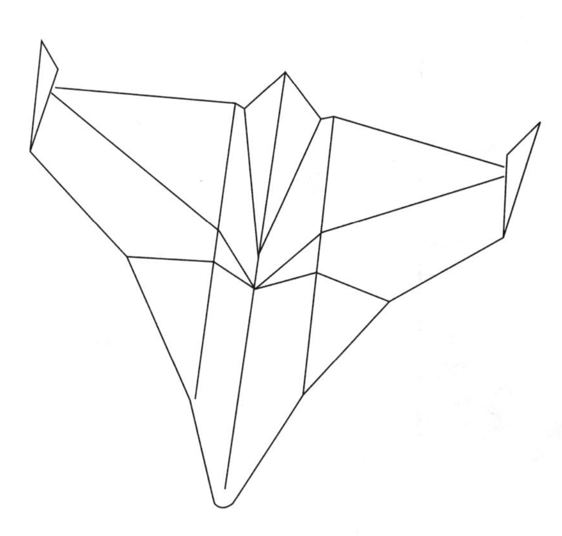

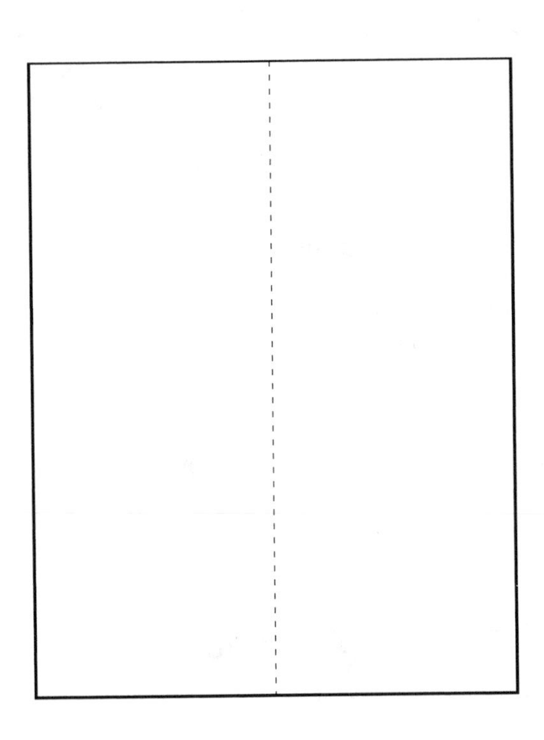

1. Make a central crease, as shown, by aligning one long edge of the paper with the edge opposite. Crease well.

Unfold.

Turn the paper over.

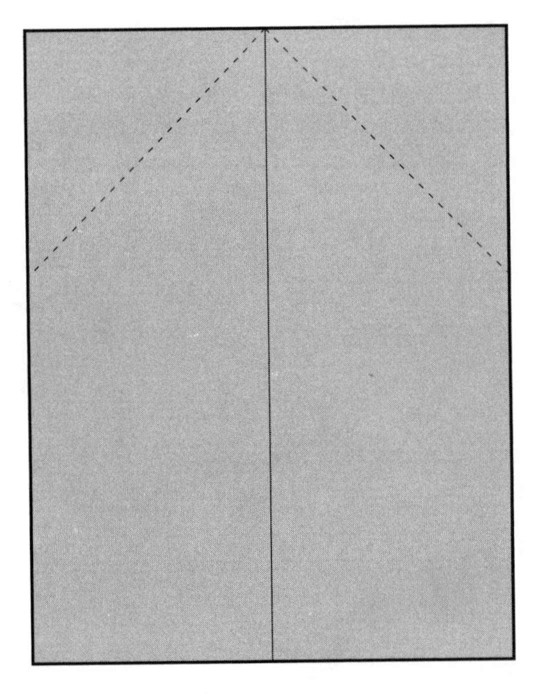

2. Fold down each top corner as shown, aligning each top edge against the central crease, as shown in Fig. 3.

3. Now fold down along the lower edge of the folds just made, as shown. Be certain to keep the central crease aligned.

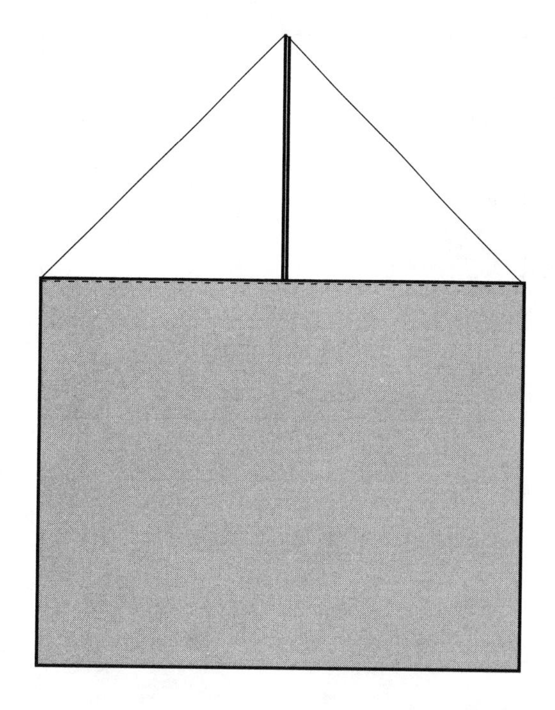

4. Shows the result of Step 3. Turn the paper over, keeping the folded edge at the top.

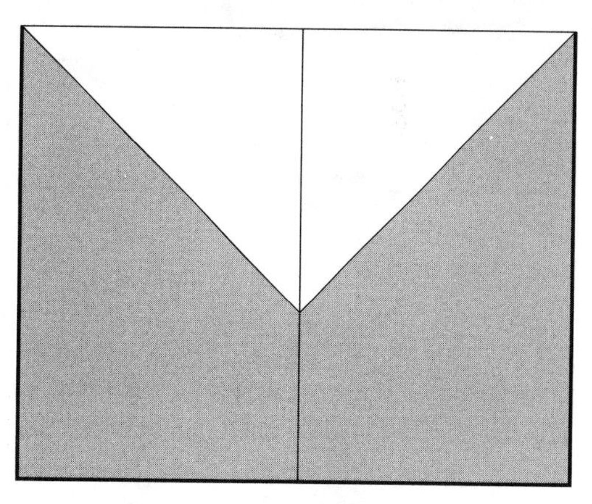

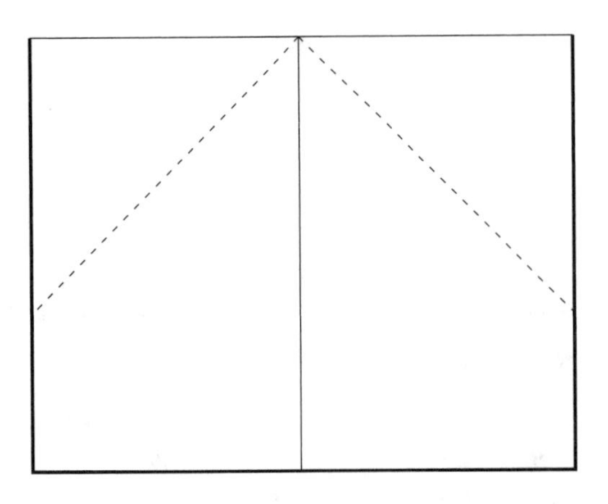

5. Making certain that the folded edge is upward, fold down both top corners as shown, aligning the top edge on each side with the central crease, as shown in Fig. 6.

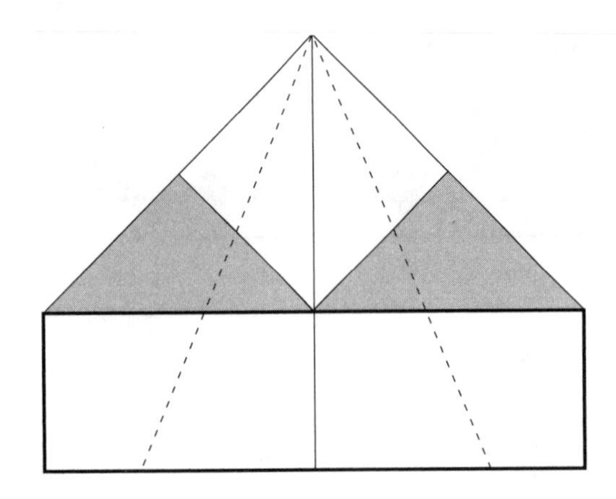

6. Make another similar fold as shown, again aligning with the central crease. The result should look like Fig. 7.

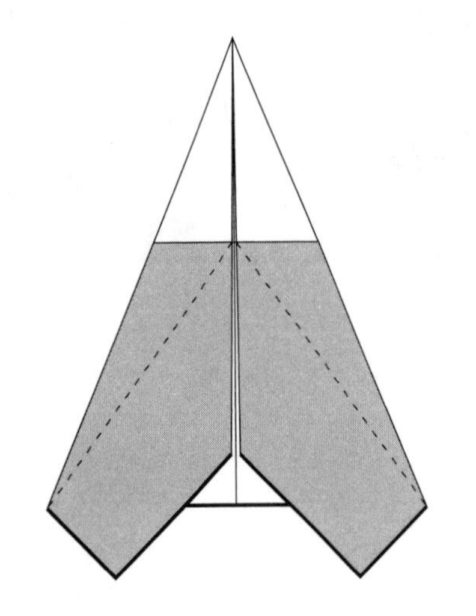

7. Now fold out each wing as shown. The fold should extend from the center of the horizontal edge toward the front, to the back corner on each side, as shown.

8. Shows the result of Step 7. Turn the plane over.

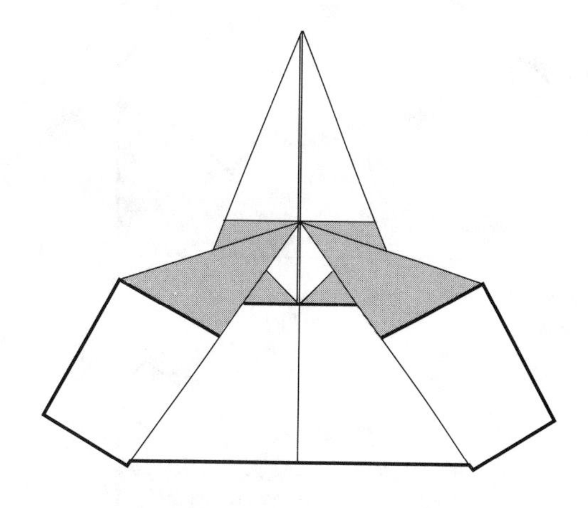

9. Fold down each wing as shown. Start by folding over the edge at point A in the front, folding so that edge BC crosses exactly over corner D in back. Fold exactly the same on each side.

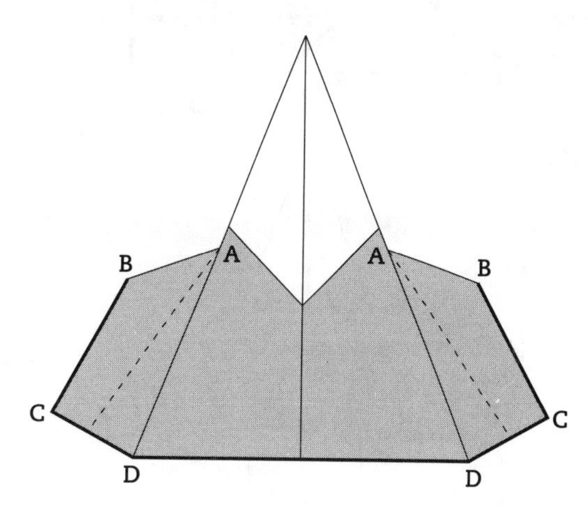

10. Shows the result of Step 9. Turn the plane over.

Fold in half along the central crease. The two wings should align exactly.

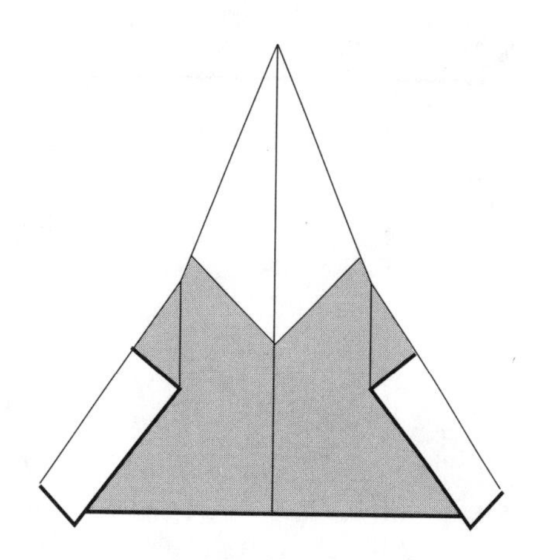

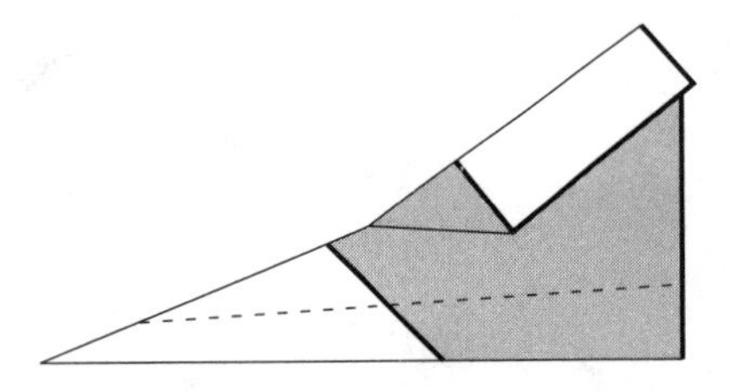

11. Fold down each wing—the fold should taper from near the central fold in front, to approximately ¾ inch from the crease in back, as shown. The exact location of the fold is not important, so long as both wings are folded exactly the same.

Unfold the wings.

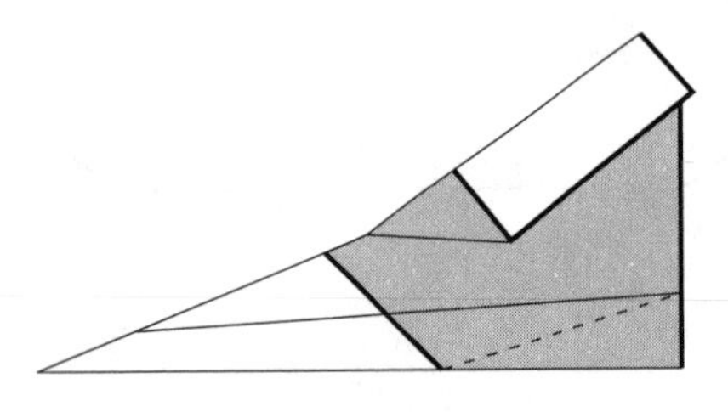

12. Now fold up the tail in an accordion fold as shown, pulling the tip of the tail upward between the wings and creasing so that the central crease of the tail is inside out. (*See also* Fig. 13.)

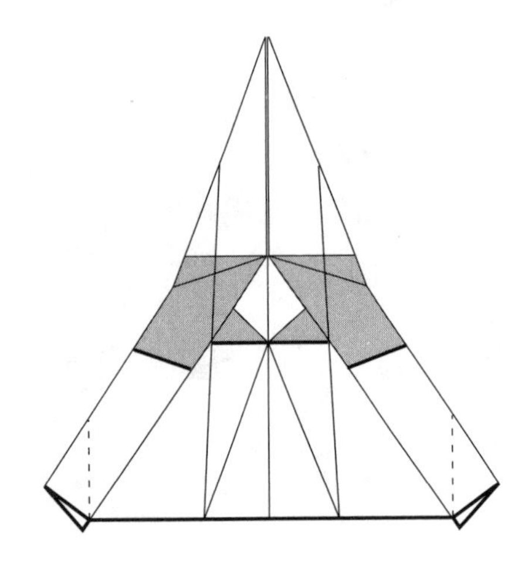

13. This is a top view showing the tail fold. Fold up the tips of the wings as shown, approximately parallel to the wing folds.

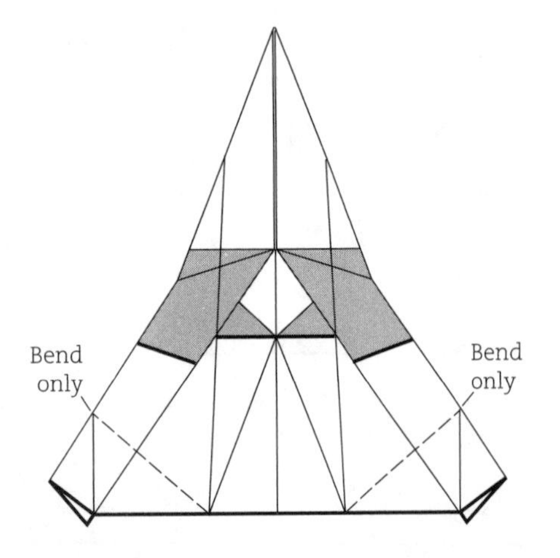

14. Gently bend up the back of each wing in approximately the area shown. Be careful not to crease.

Flying instructions for PLANE 15 are on page 81.

Navy jet

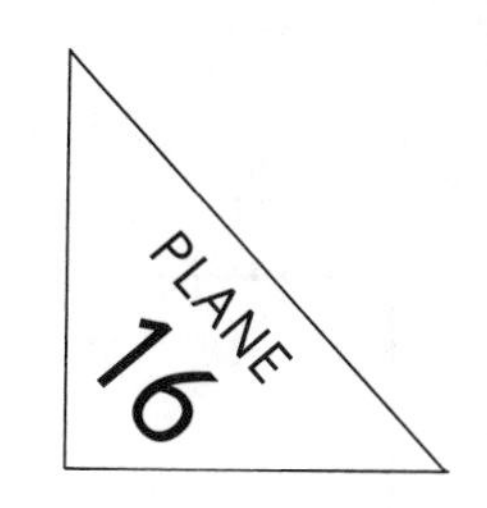

A PERSONAL FAVORITE: This plane is modeled after the fighter jets that used to roar overhead when we lived off the runway of a U.S. Navy air base, and is designed intentionally nose-heavy for faster speed. It is not hard to fold, but usually requires some adjusting and a few trial throws before you get it flying well. Skillfully thrown, it will perform loops and other aerobatic stunts, and the heavily folded nose acts as a shock absorber, making this plane nearly indestructible. (Folding instructions are on pages 88–91.)

To fly

Hold from beneath and throw. Start with gentle tosses until well-adjusted, then take outside and try throwing harder, either straight ahead or upward. If there is a breeze, throw into the wind. Continue adjusting as needed.

Adjustments

Sight from behind before you make your first throw, making certain that the wings and folds match on each side. This jet always requires a large elevator, hence Step 11 of the folding instructions. Make short test flights, adjusting the elevators as described in the introduction subsection entitled "Making it fly," until the flight is straight and even. You are then ready for harder tosses. By adding more elevator on each side, you can make the plane do a large loop, then swoop down for a smooth landing. If it dives into the ground instead of landing, simply straighten the nose and bend the elevators up farther on each side. Good flights require continuing adjustment, and this takes practice.

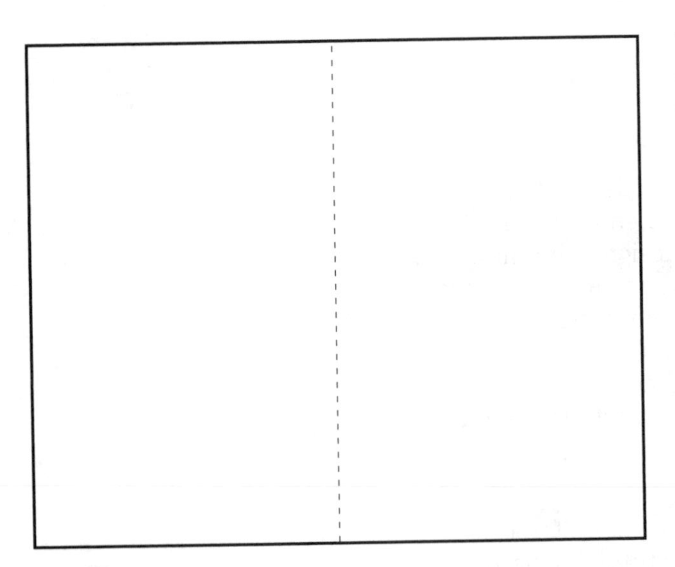

1. With the paper as shown, make a central crease by aligning the left edge of the paper against the right. Crease well.

Unfold.

Turn the paper over.

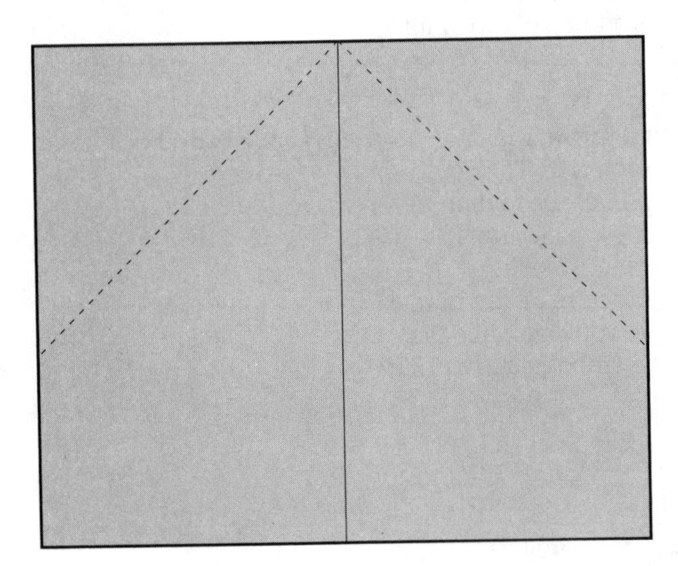

2. Fold down both top corners as shown, aligning the top edge on each side with the central crease.

3. Shows the result of Step 2. Fold down the tip as shown, so that the tip rests exactly on the center where the horizontal paper edges meet, as shown in Fig. 4. Crease well.

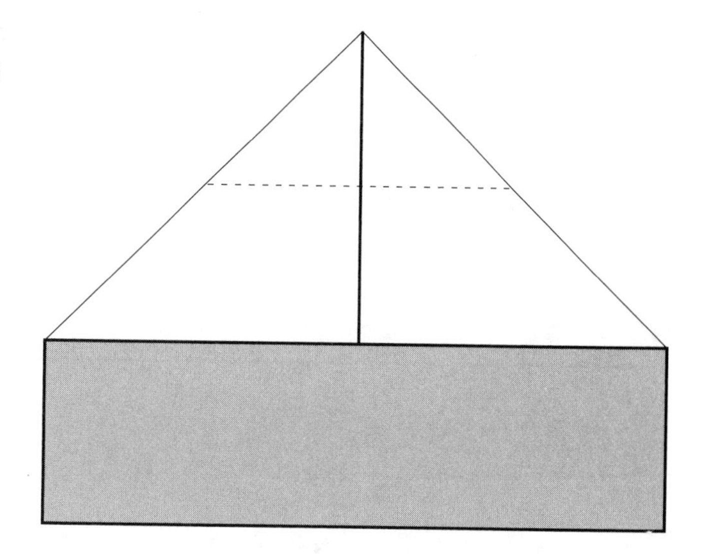

4. Now fold down the top corners as shown, again aligning the top edge against the central crease.

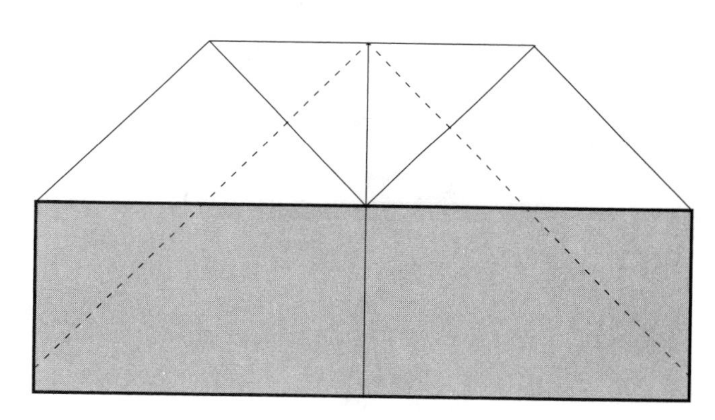

5. This is the result of Step 4. Turn the plane over.

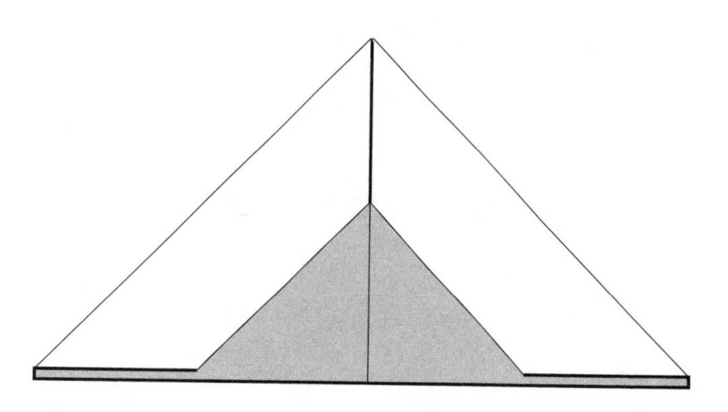

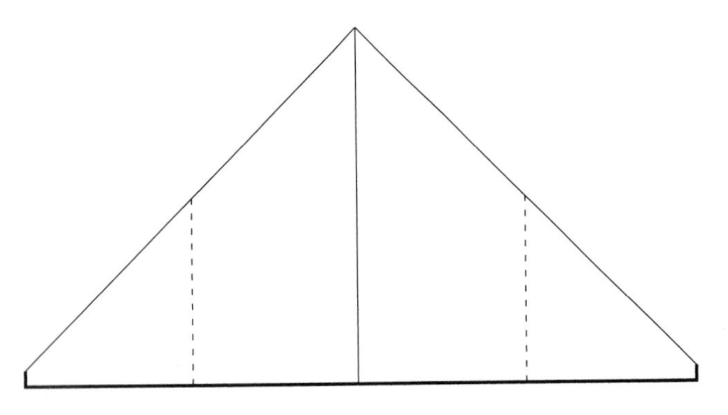

6. Fold each wingtip toward the center so that they exactly touch the central crease in the back. Keep the back of the wing perfectly aligned. The result should look like Fig. 7.

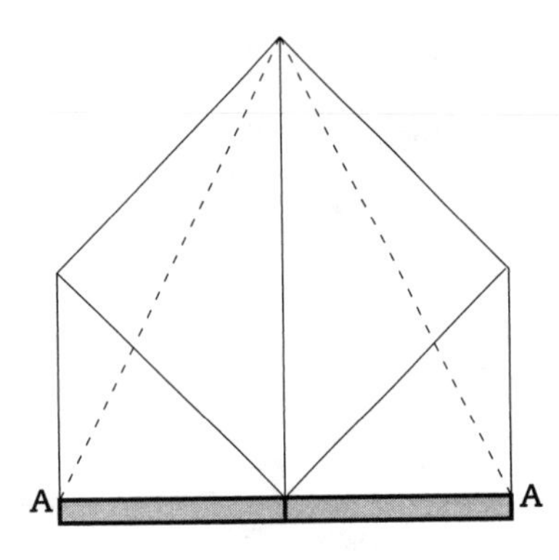

7. Make a diagonal fold as shown from the front tip of the plane to point A on each side, to form the front edge of the wings. Crease very well.

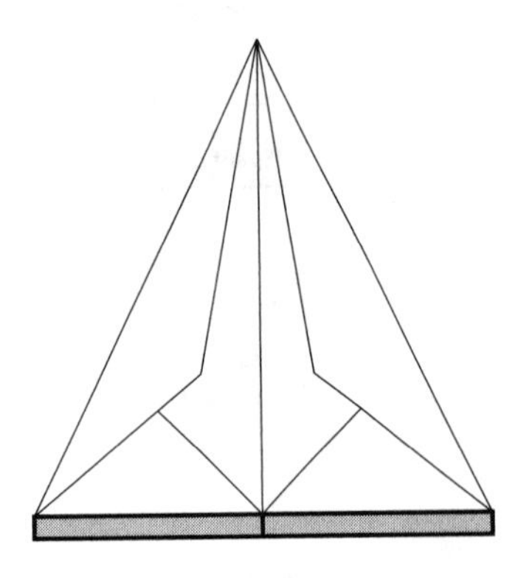

8. Shows the result of Step 7. Now fold the plane in half along the central crease.

9. Fold down each wing as shown. The fold should start a little over an inch from the front of the plane, and should taper toward the central fold in the back, as in the illustration. The wings should be aligned perfectly with each other—the exact location of the fold is less important than making it the same on each side.

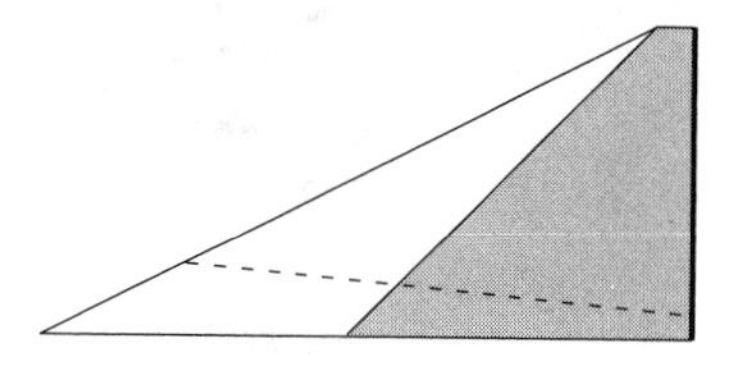

10. This drawing shows the plane from the side, with the wing folded down. On each side, fold out the tail as shown, making the fold approximately parallel to the wing fold from Step 9. This fold should be as equal as possible on each side.

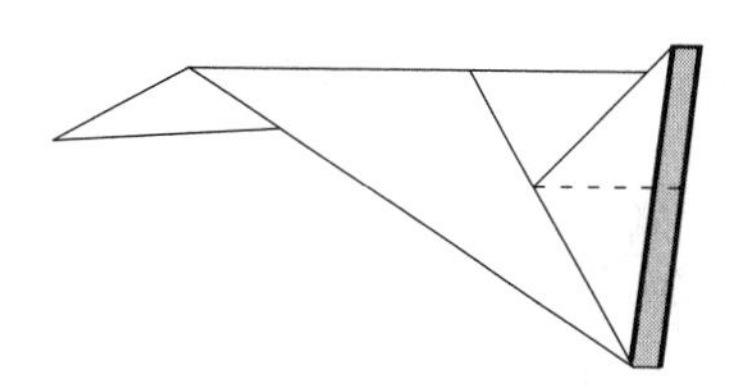

11. This is a top view. Gently fold up the outer tip of each wing as shown, to make an elevator on each side.

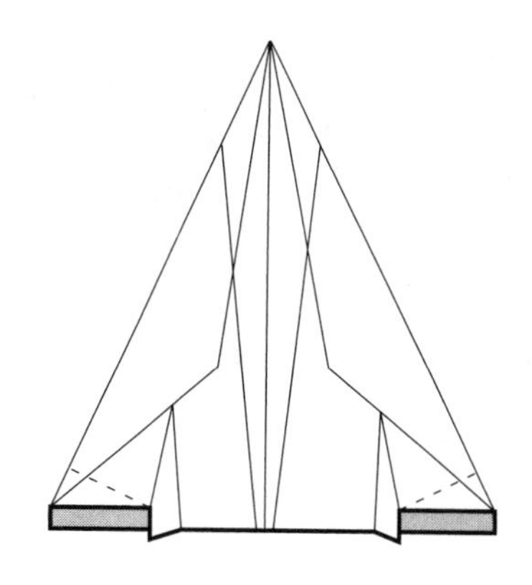

Flying instructions for PLANE 16 are on page 87.

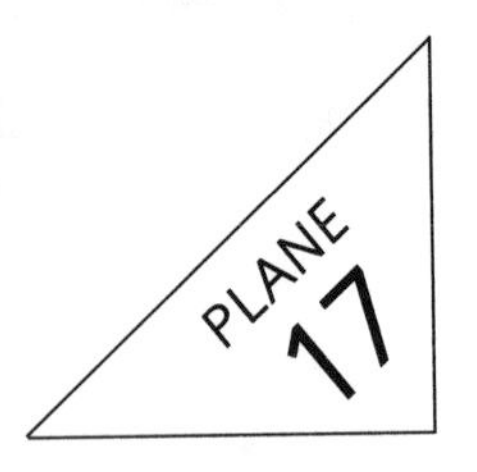

Canard jet

A CANARD is a small wing placed on an aircraft forward of the main wing, such as was found on the original Wright brothers' flying machine. Largely abandoned for decades, the canard has begun to find its way back into modern aircraft design. This canard model is fairly easy to fold and fly; for a more challenging design, *see* PLANE 20. (Folding instructions for PLANE 20 are on pages 112–119.)

To fly

Hold from beneath and throw forward.

Adjustments

This plane rarely requires adjustment to fly well. It is interesting, however, to make adjustments on the canard wing rather than on the main wing, as you will find that the adjustments have the opposite effect to those described in the introduction's subsection making it fly. To make the plane climb, for example, you can bend down the back edge of both canards, and to make it nose down, bend the same edges upward. (Folding instructions are on pages 93–97.)

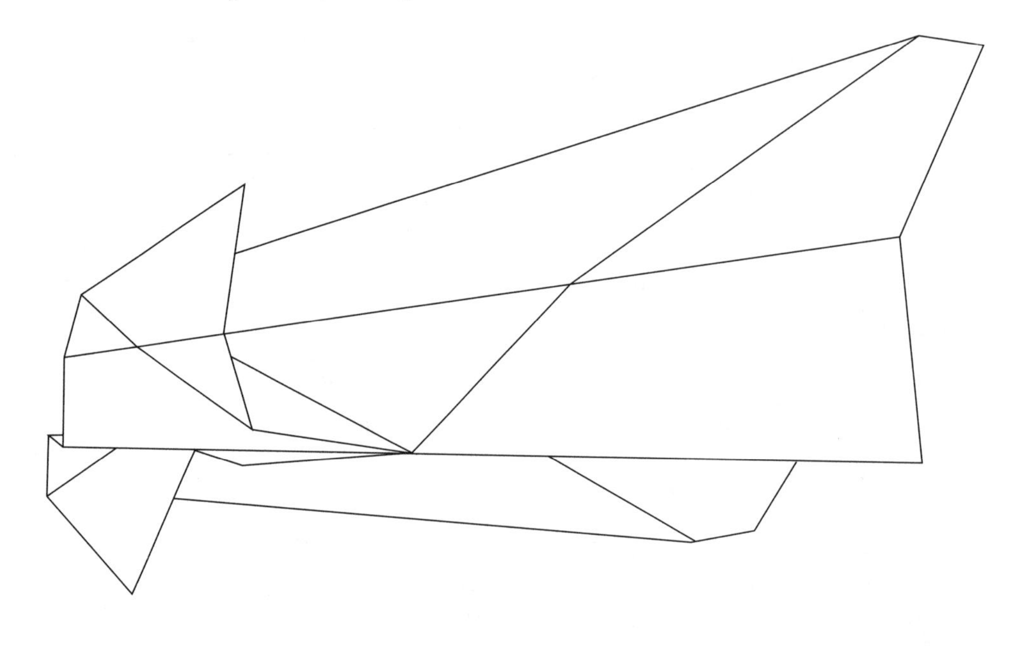

1. Make a central crease as shown, by aligning one long edge of the paper with the opposite long edge. Crease well.

Unfold.

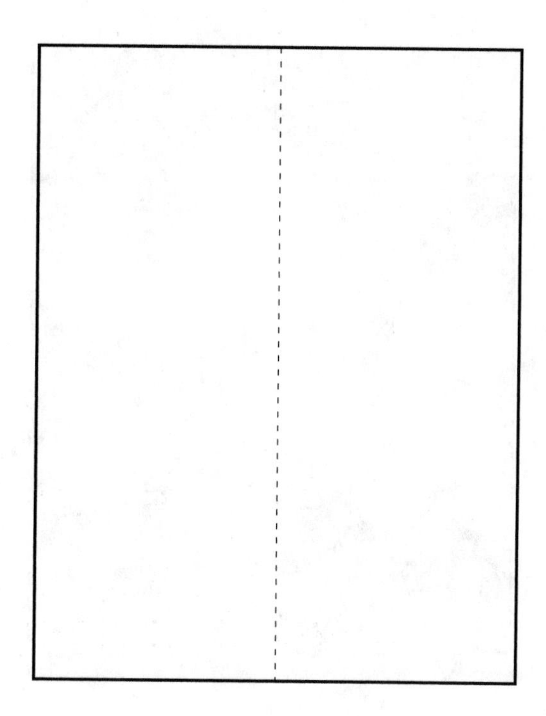

2. Fold down both top corners as shown, aligning the top edge on each side against the central crease. Crease well.

Unfold both sides.

Turn the paper over.

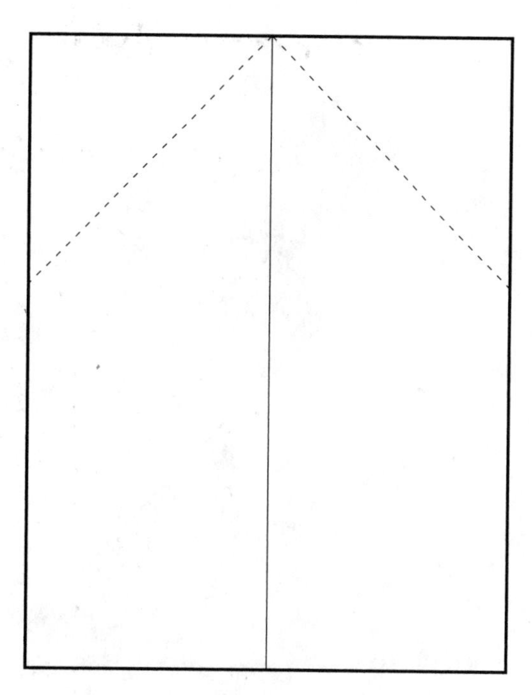

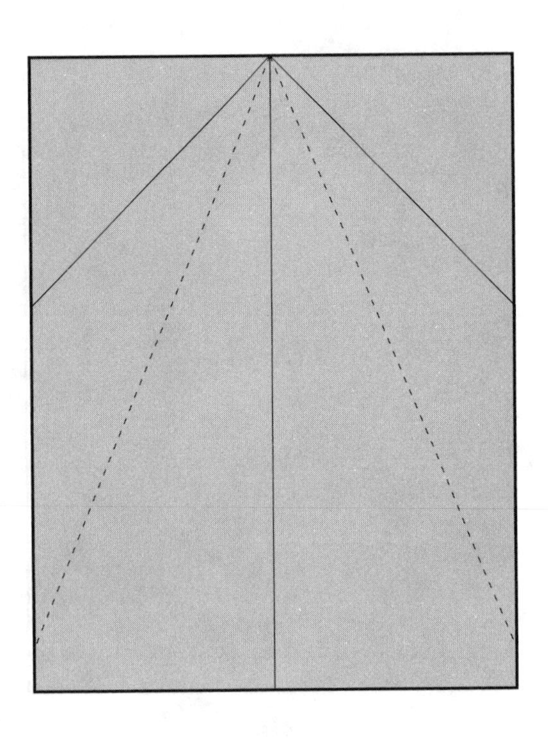

3. Note that the creases you just made join each other at the front of the central crease. Take the diagonal crease on one side and align it carefully against the central crease, and fold as shown. Repeat on the other side. The result should look like Fig. 4.

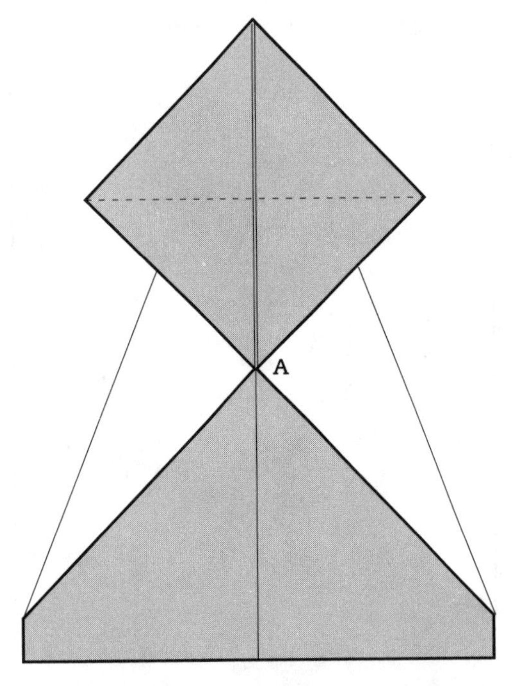

4. Now fold back the front tip so that it touches point A. Crease well.

5. Shows the result of the previous fold. Fold the tip back forward so that it just touches the front edge in the exact center.

Unfold.

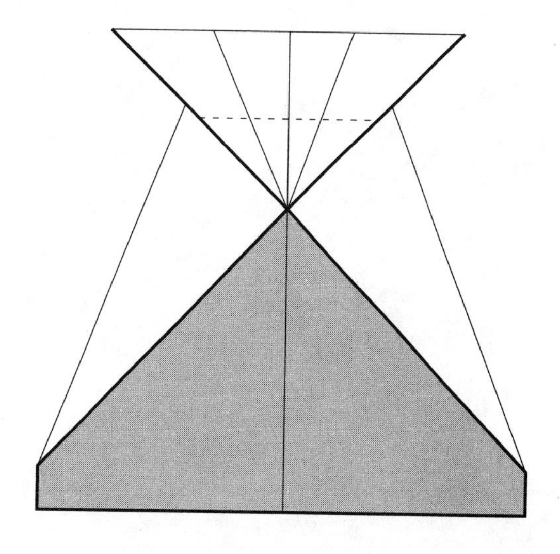

6. Fold up the small forward wing on the right, following the paper edge shown.

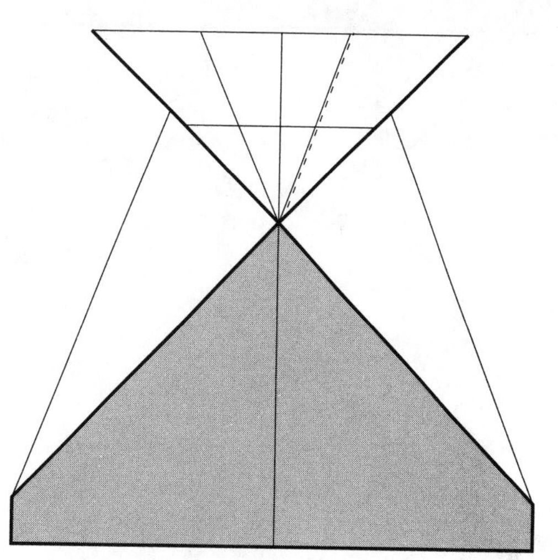

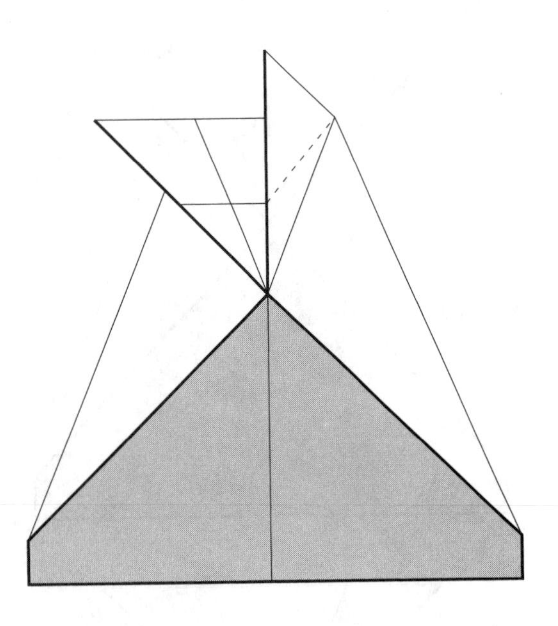

7. Now fold the small right wing back out as shown. Your fold should extend from the right upper corner to the horizontal crease in the center, as shown.

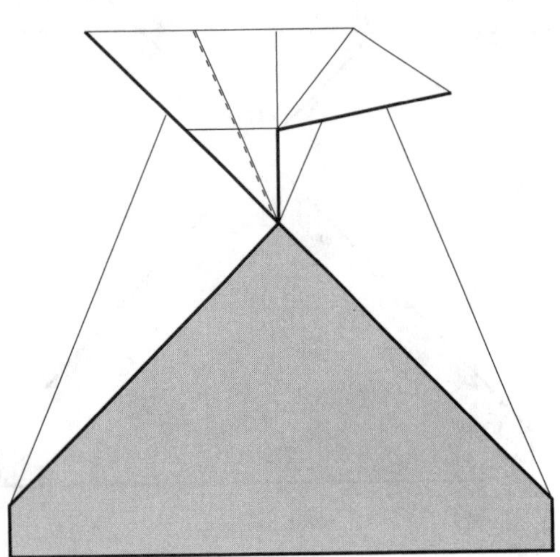

8. Shows the result of Step 7. Now, fold up the small left wing, as in Step 6.

9. Fold down the left wing, as in Step 7.

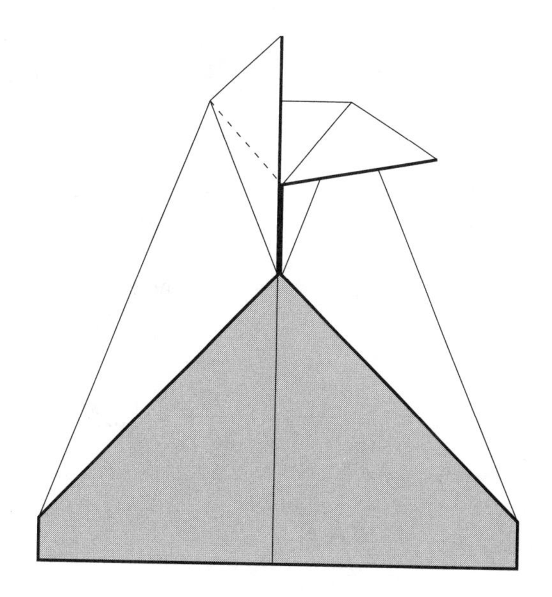

10. The result should look like this. Turn the plane over.

Fold in half along the central crease, carefully aligning both sets of wings.

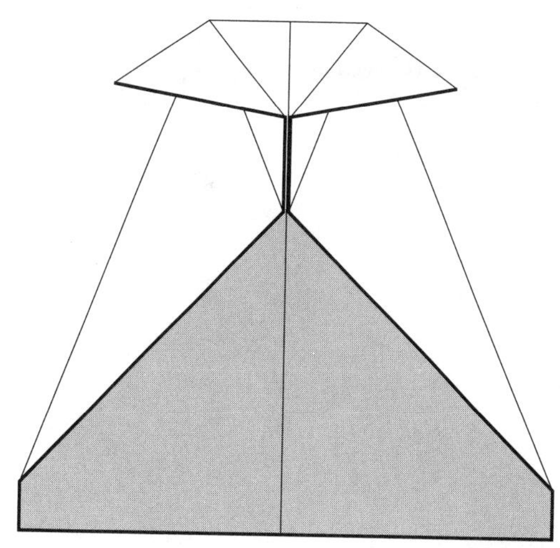

11. Fold down each wing as shown, aligning the diagonal edge of each wing with the central fold. Make certain both the wings and the canards match each other exactly. Partially unfold.

Flying instructions for PLANE 17 are on page 92.

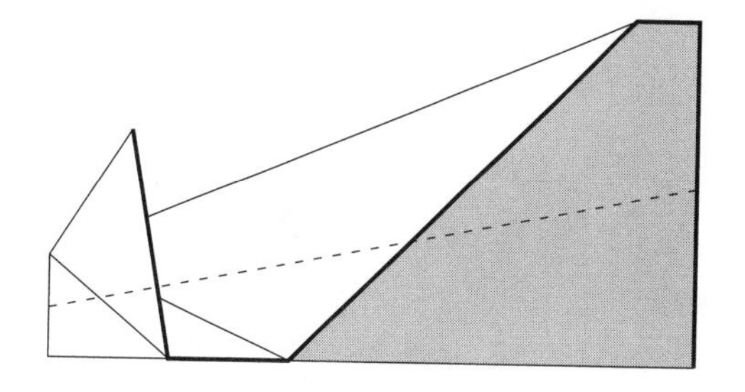

Stealth flying wing

ANOTHER CONCEPT toyed with from time to time by aeronautical engineers is the flying wing—that is, a plane that is all wing, with no body or tail. Such is the design of the controversial stealth bomber that was made public not many years ago, and of which this paper plane is a little reminiscent. Like its real life counterpart, this plane flies remarkably well and is virtually invisible to radar—yet the cost (on paper at least) is quite modest. (Folding instructions are on pages 99–104.)

To fly

Hold the flying wing from behind, shove forward horizontally, and release.

Adjustments

Make certain the front edge of the wing is well creased. Notice that the wing opens up in the rear like a pocket-bread sandwich. The key to adjusting for a straight, even flight is widening or flattening this opening and making it the same on both sides. (Widening just one side will make the plane turn toward that side.) Usually little adjustment is needed.

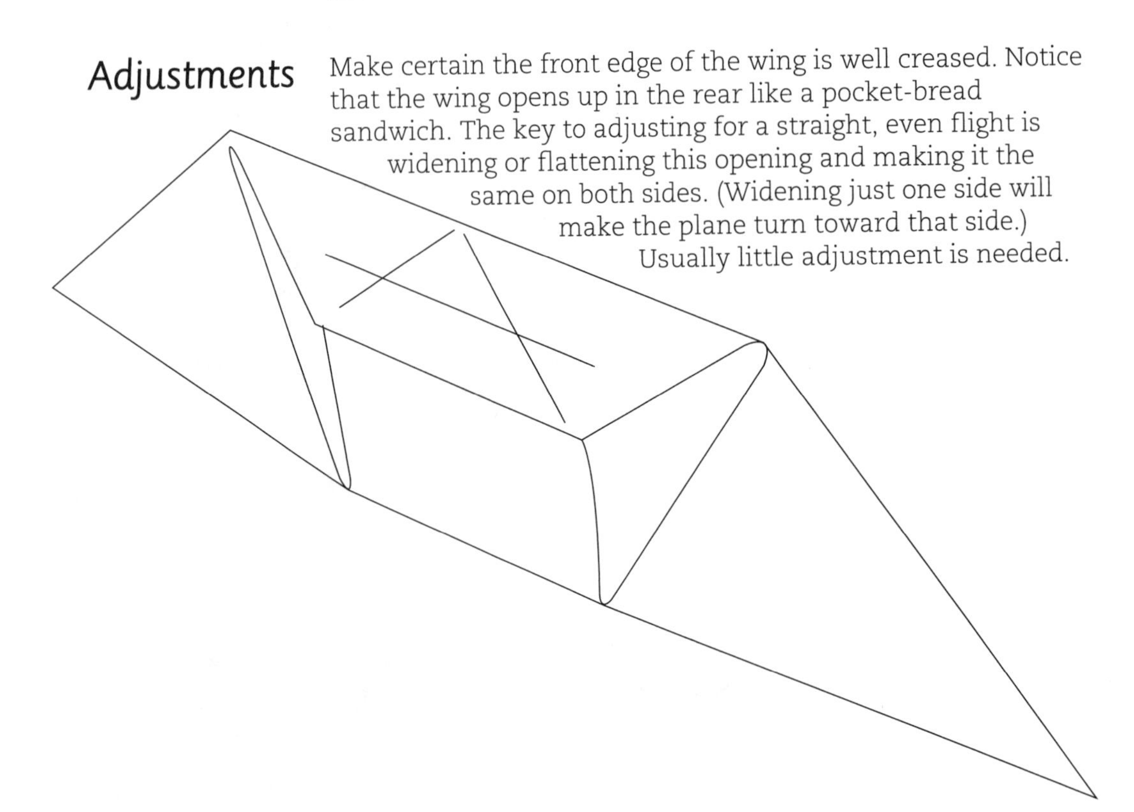

1. Make a fold from one corner by aligning the side edge against the top edge, as shown. Crease well.

Unfold.

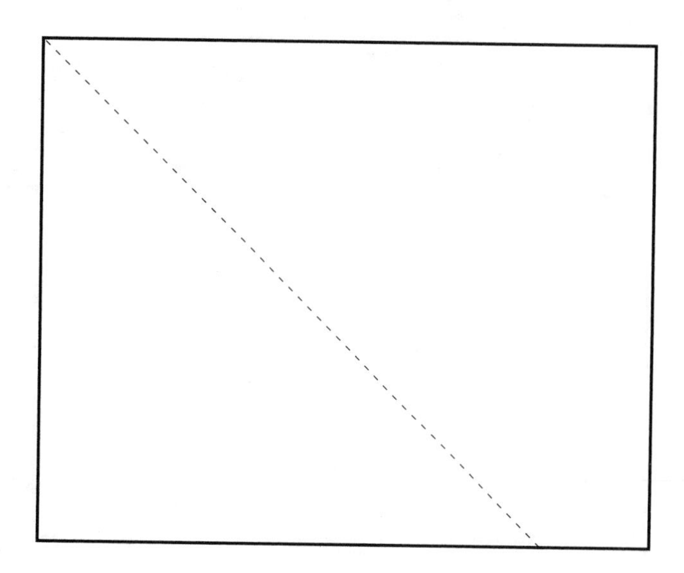

2. Make a similar fold on the bottom corner opposite, as shown. Crease well.

Unfold.

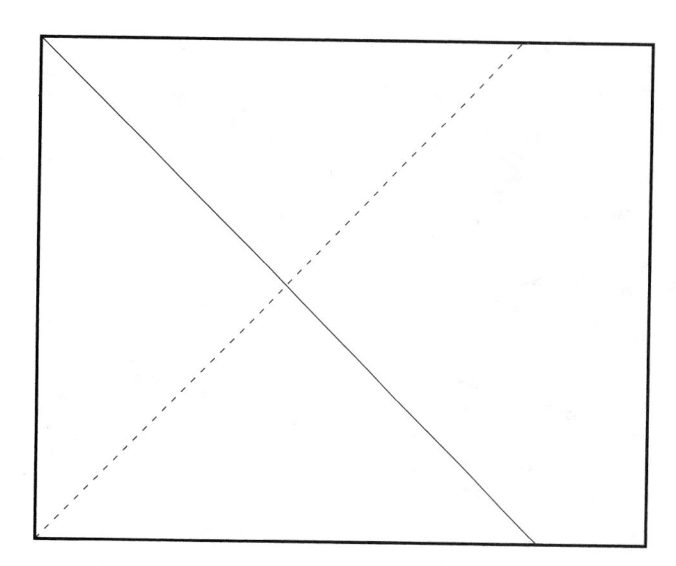

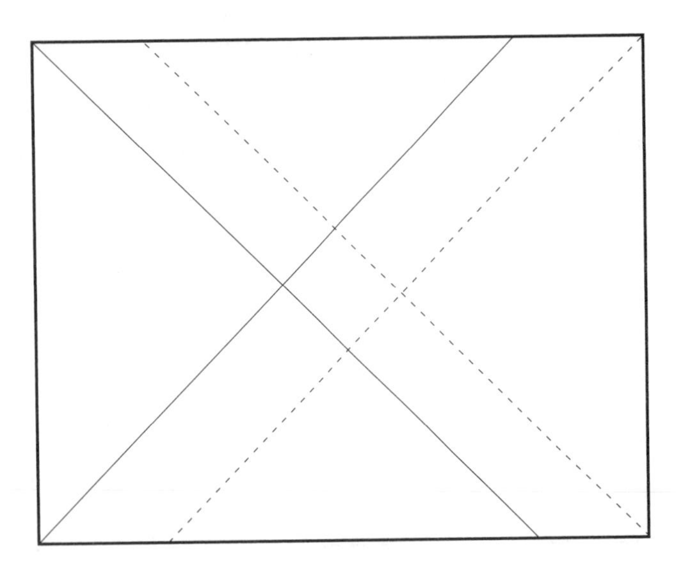

3. Make similar folds from the upper and lower corners of the opposite edge of the paper as shown. Crease each well.

Unfold each in turn.

Turn the paper over.

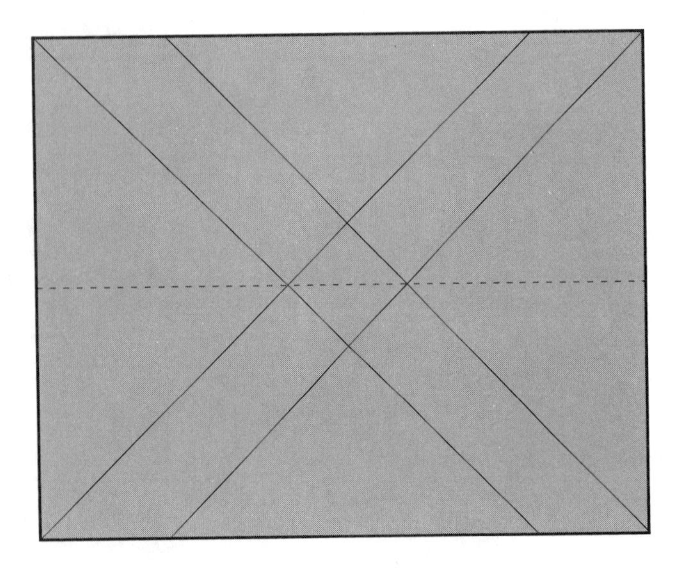

4. Now make a horizontal central crease as shown, by aligning the bottom edge to the top.

Unfold.

Turn the paper over.

5. Now you will make a complex accordion fold, as follows:

- ~ Push in at point A and B with your finger, which should start the accordion fold.
- ~ Pull points C and D toward the bottom of the paper; the paper should fold along the lower diagonal creases extending from corners E and F.
- ~ Turn inside out the small section of central crease between point A and B so the paper can lay flat.
- ~ Bring down corners G and H so that they lay atop corners E and F. The diagonal folds should match each other exactly.

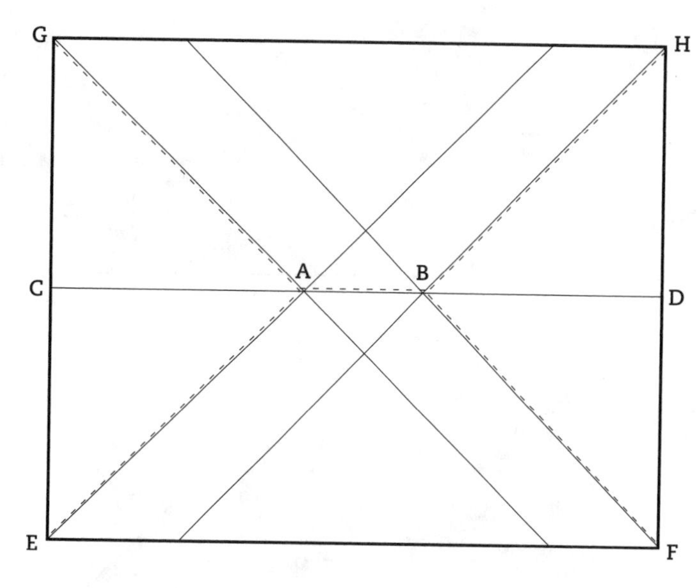

6. This shows the fold in progress. Carefully recrease all folds, keeping the upper and lower wings perfectly aligned with each other.

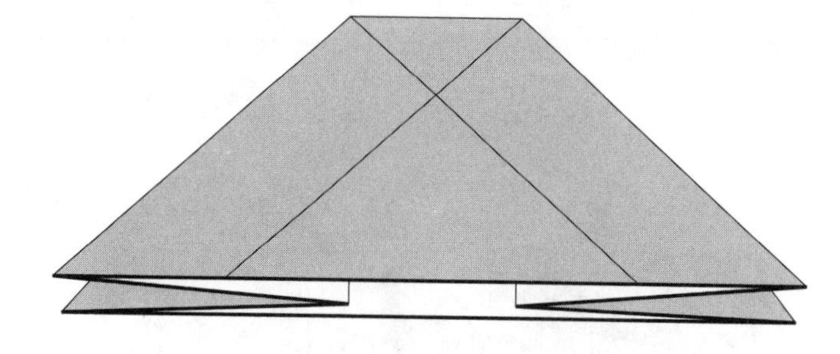

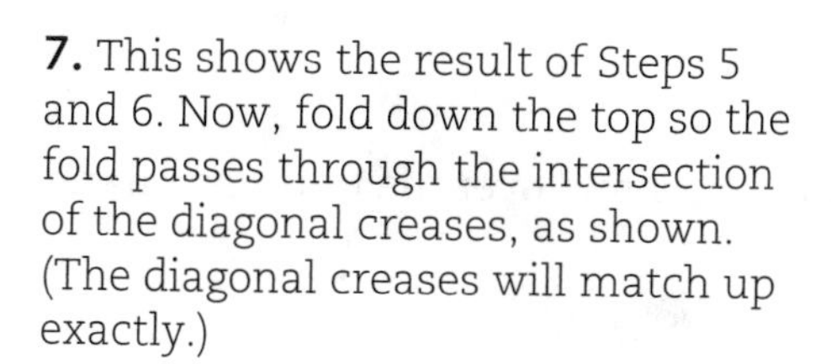

7. This shows the result of Steps 5 and 6. Now, fold down the top so the fold passes through the intersection of the diagonal creases, as shown. (The diagonal creases will match up exactly.)

Crease well.

Unfold.

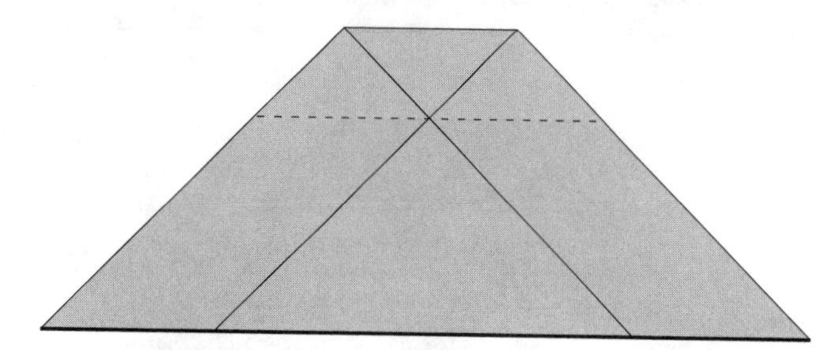

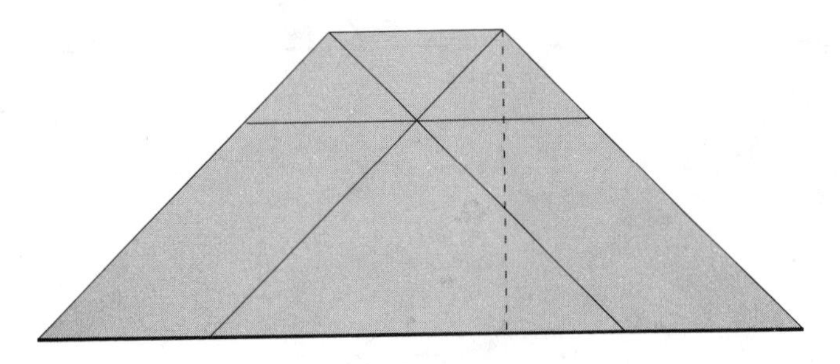

8. Note that each side of the plane consists of two "wings." Fold over the uppermost of these on the right as shown, keeping the fold straight by aligning the bottom edges. The result will be as in Fig. 9.

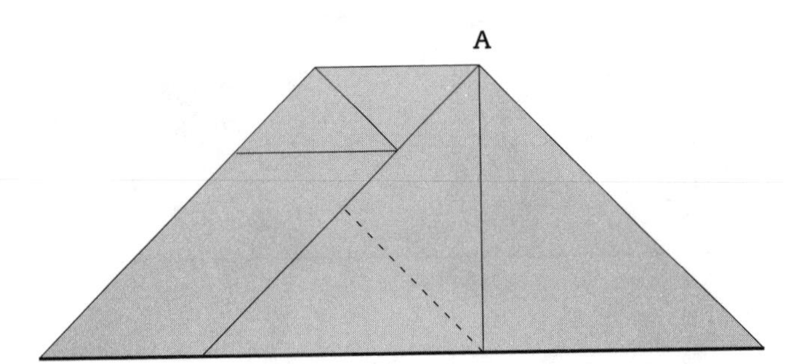

9. Now, fold up the bottom tip against the top corner A as shown.

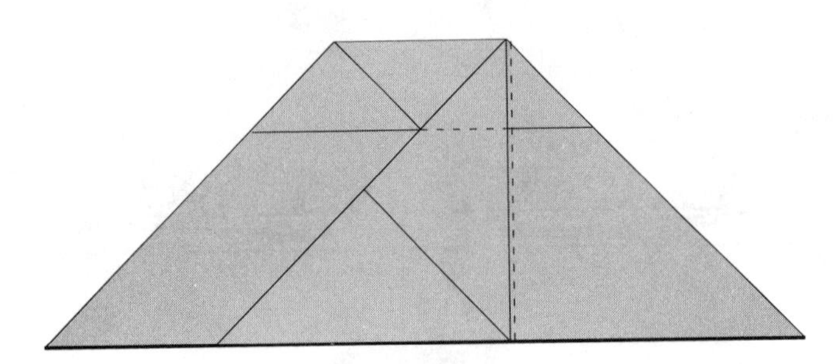

10. Fold the same tip down, using as a guide the horizontal crease you formed in Step 7.

Fold the entire right wing back to its original position.

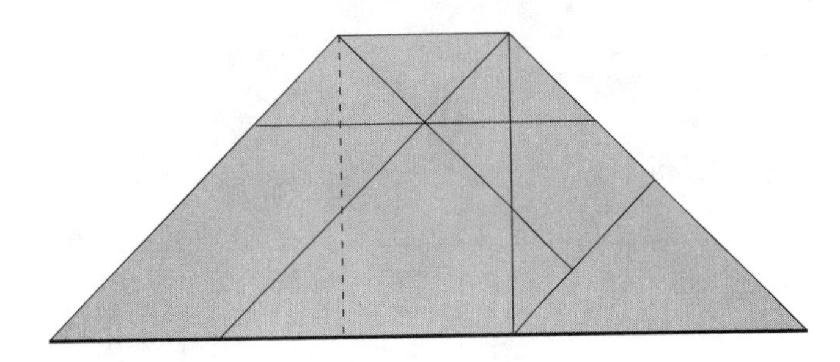

11. Now fold over the top wing on the left.

12. Fold up the tip on the left as you did on the right.

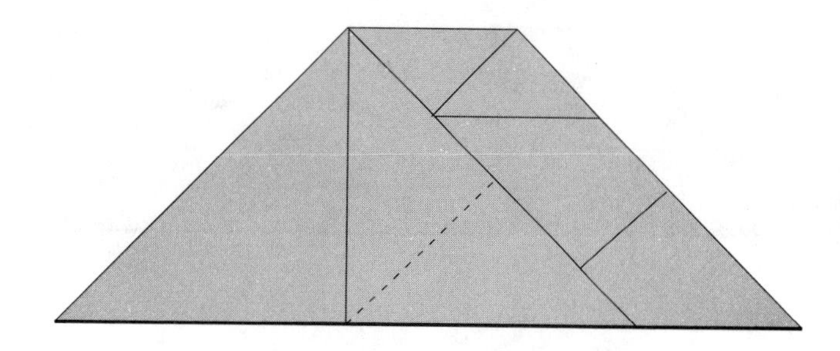

13. Fold down the tip following the horizontal crease, as you did on the right.

Fold the left wing back to its original position.

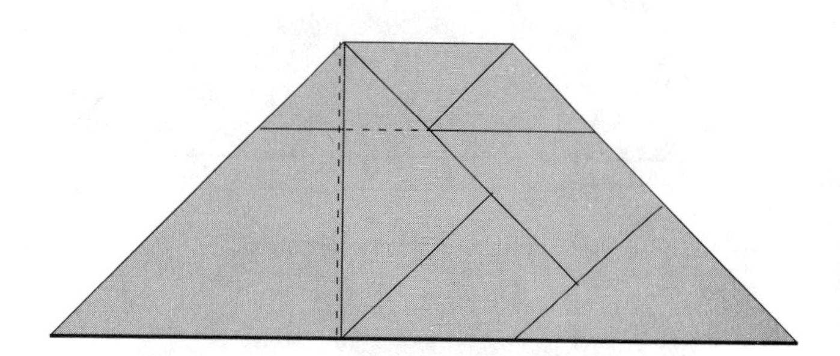

14. Fold down the top edge so it lines up with the horizontal crease you formed in Step 7. Then unfold.

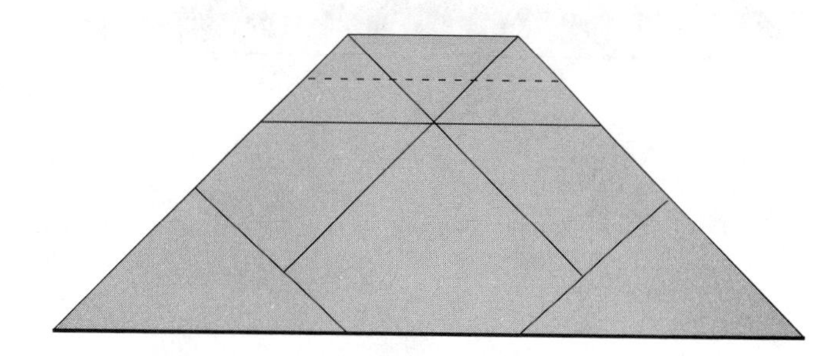

15. Fold down along the horizontal crease.

Note that you have created a little pocket in each side of the folded-down section. (*See* Step 16.) Open this pocket gently on each side with your finger.

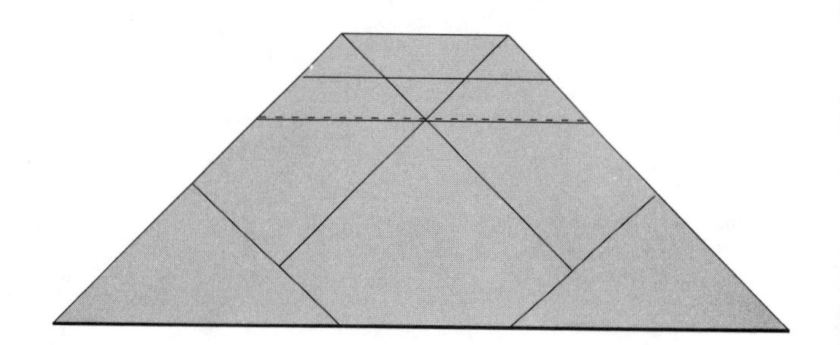

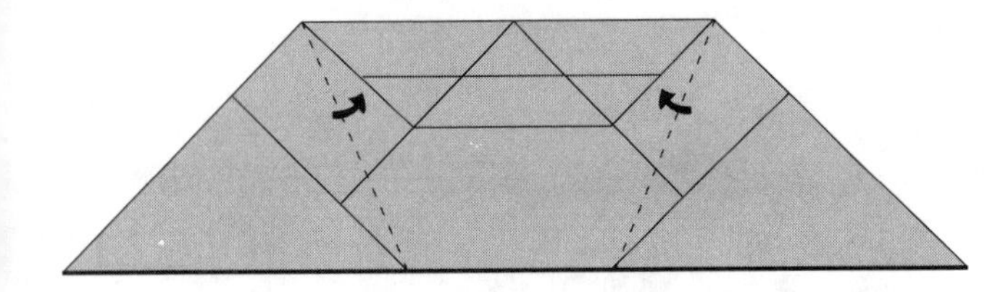

16. Now fold up the remaining side flaps as shown on each side. Once folded, tuck the top corner of each into the side pockets created in Step 15.

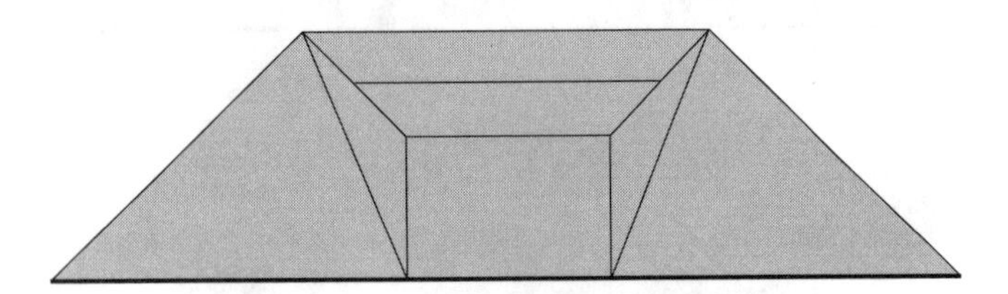

17. This shows the result of Step 16. Turn the plane over.

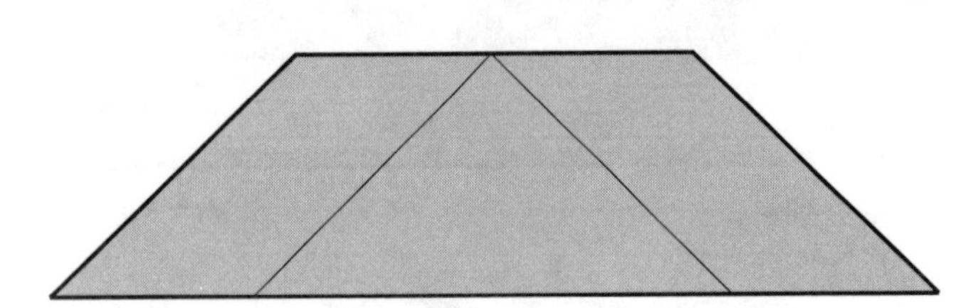

18. The finished plane, from above.

Flying instructions for PLANE 18 are on page 98.

Batwing

PLANE 19

ANOTHER VARIATION of the flying wing concept, this plane is strictly for indoor flight. It is less difficult but less predictable than PLANE 18 and has something of a wild look to it. (Folding instructions are on pages 106–110.)

To fly

Place your index fingertip over the back-pointing tail, with your other fingers beneath. Push smoothly, directly forward, and release.

Adjustments

The key to this plane is symmetry. Make sure that the wings are evenly matched and equally flattened, and that they angle the same. If the plane deviates to one side, unevenness is invariably the problem. If the plane glides sluggishly, refold along the central crease to increase the upslope of the wings. It sometimes helps to exaggerate this upward fold at the tail with your fingers. When well-adjusted, the plane almost leaps forward in a smooth, even flight.

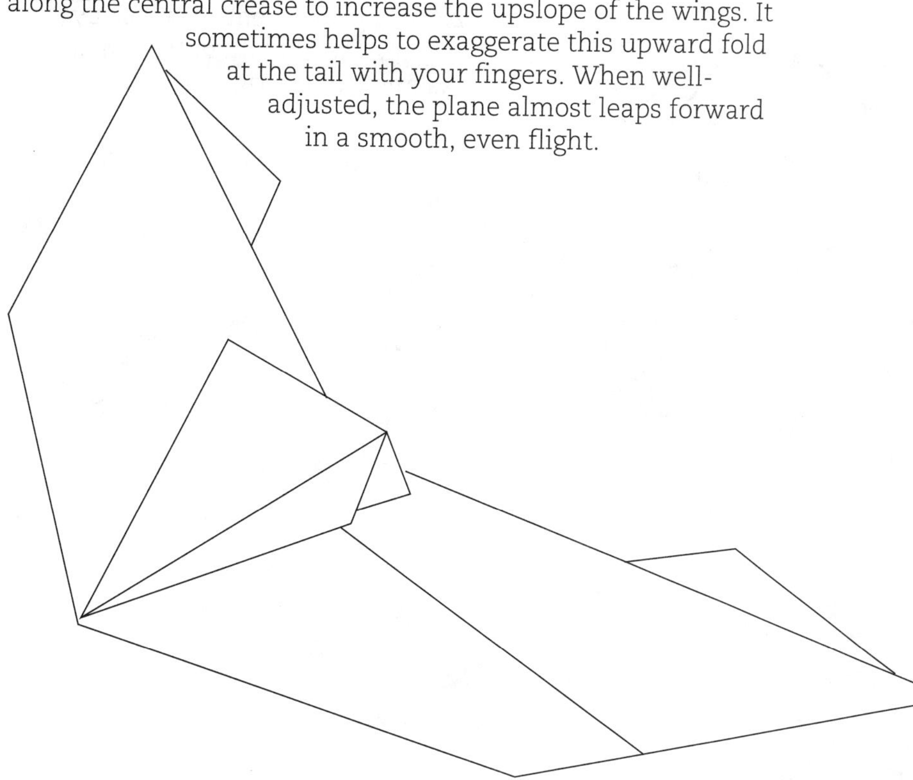

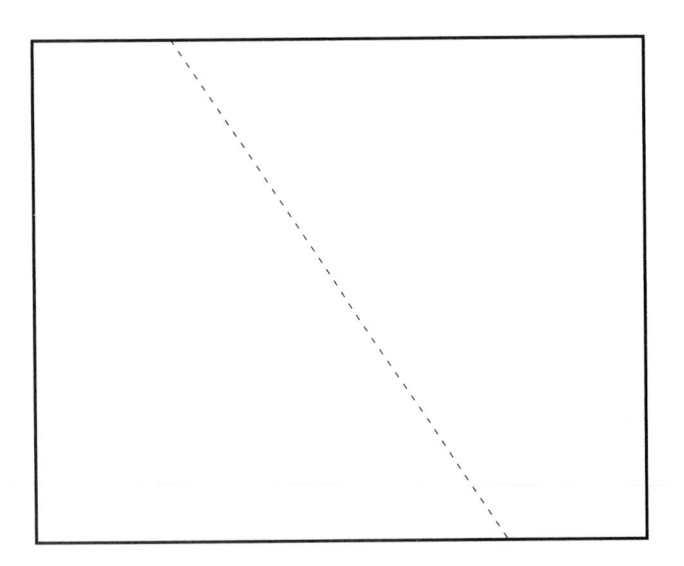

1. Make a diagonal fold as shown, by aligning the left bottom corner with the right top corner. Crease well. Place the paper as in Fig. 2.

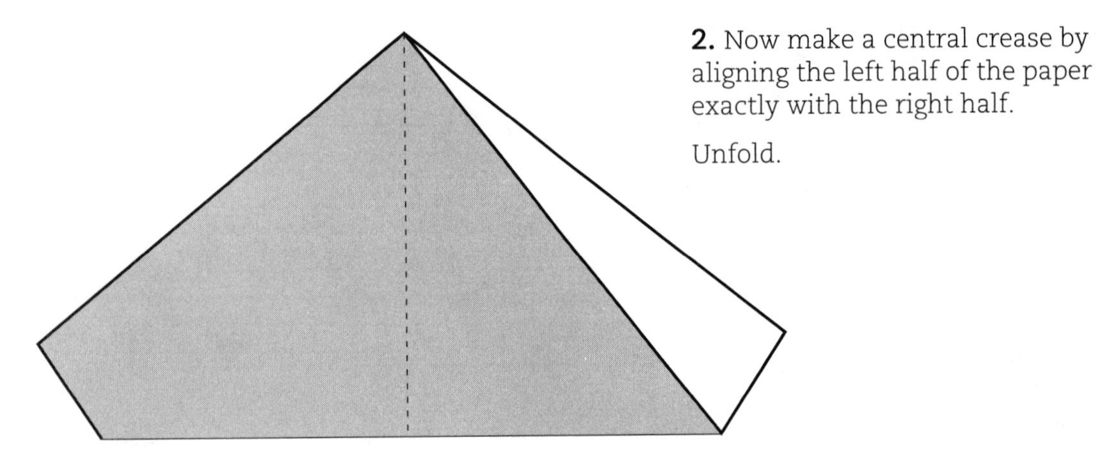

2. Now make a central crease by aligning the left half of the paper exactly with the right half.

Unfold.

3. Mark the exact center of the paper by bringing the top down to point A at the bottom center of the paper. Make a small crease as shown with your finger across the central crease.

Unfold.

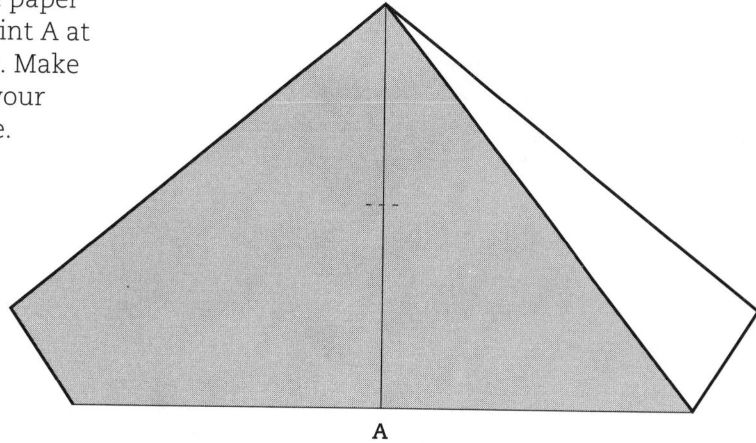

4. Now fold up the bottom edge of the paper so the edge lies flush with the mark you made in Step 3. Align the central crease, and crease well on the bottom edge. Note that your fold should pass from corner to corner as shown.

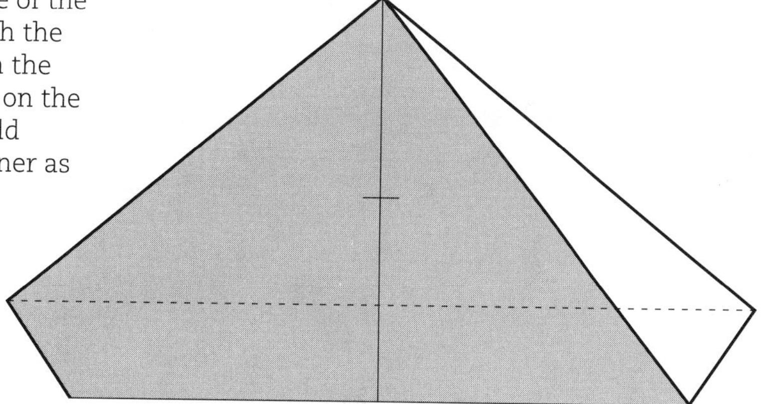

5. Make a diagonal fold on the lower portion of the paper as shown. This is done by beginning your fold outward from the point marked A, at the same time aligning point B along the bottom edge as shown in Fig. 6.

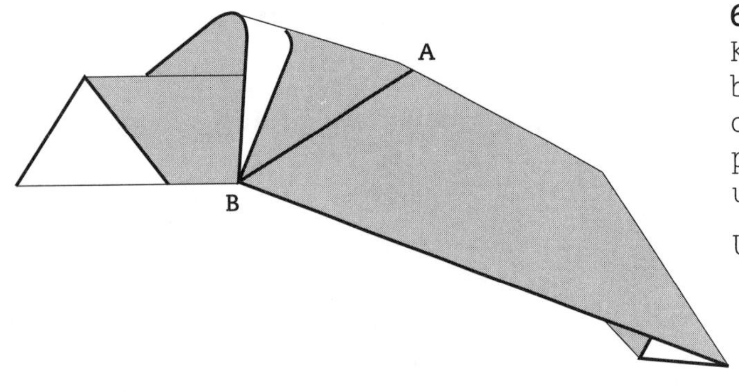

6. Shows this fold in progress. Keeping tip B pressed against the bottom edge, crease from point A outward, keeping the right side of the paper perfectly flat. Do not crease upward beyond point A.

Unfold.

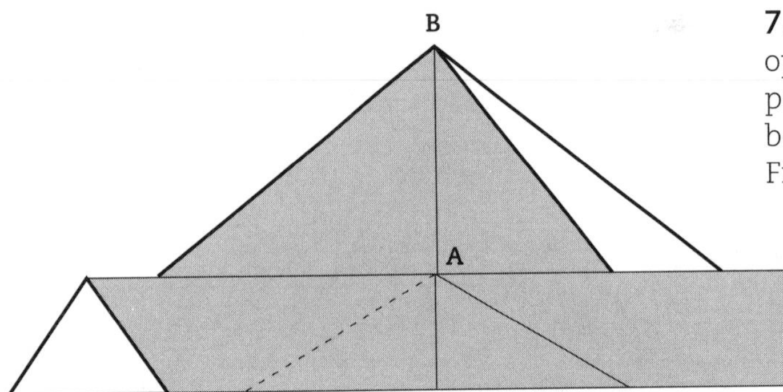

7. Now make a similar fold on the opposite side. Fold outward from point A, aligning point B with the bottom edge on the right, as shown in Fig. 8.

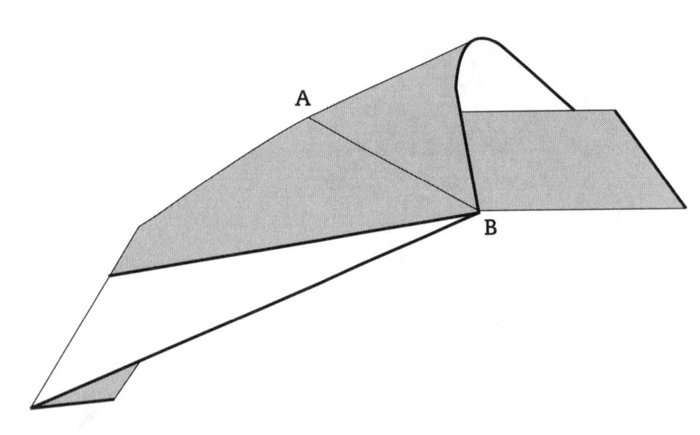

8. Shows the fold in progress. Keeping tip B pressed against the bottom edge on the right, crease from point A leftward, keeping the left side of the paper perfectly flat. Do not crease upward beyond point A.

Unfold.

9. Take both wings of the plane and fold both simultaneously downward over the diagonal folds as shown.

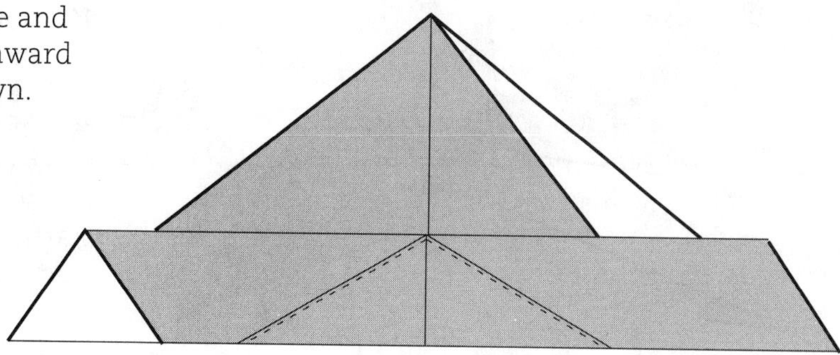

10. Shows the result. Note that there is a triangular section of paper bulging upward toward you, ending at tip B. Take this tip, and lay it over the point marked C on the right. Crease downward along the center to flatten the fold, as in Fig. 11.

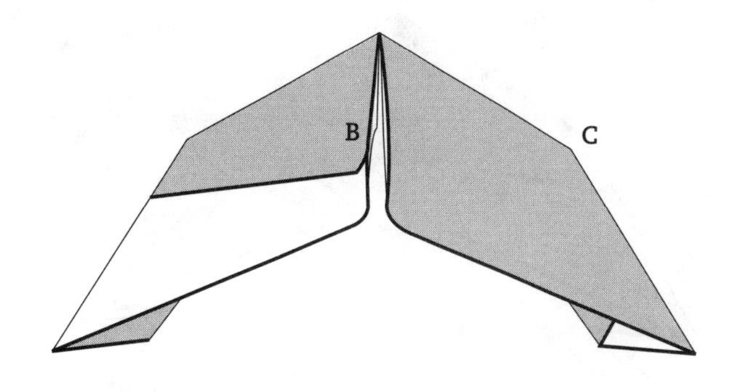

11. Note that the tip is lying to the right. Fold it over to the left as shown, laying it against point C on the left side. Crease the fold flat.

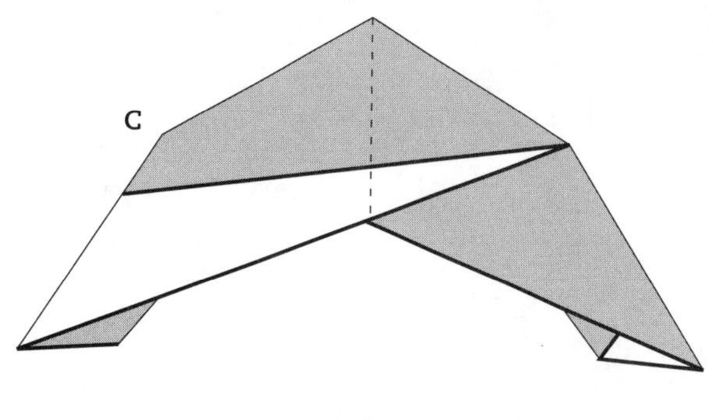

12. Shows the result. Allow the tip to point toward you again, as in Fig. 13.

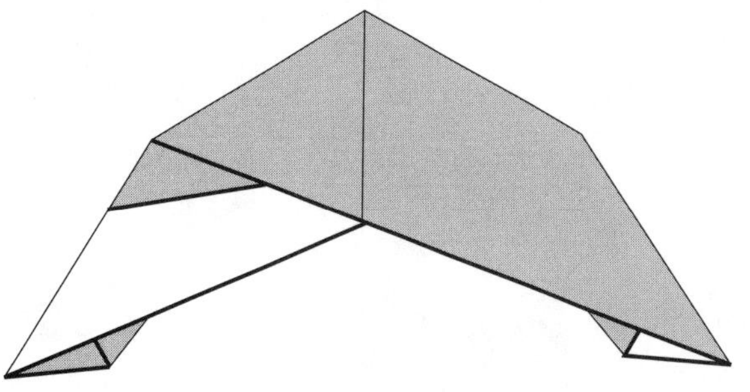

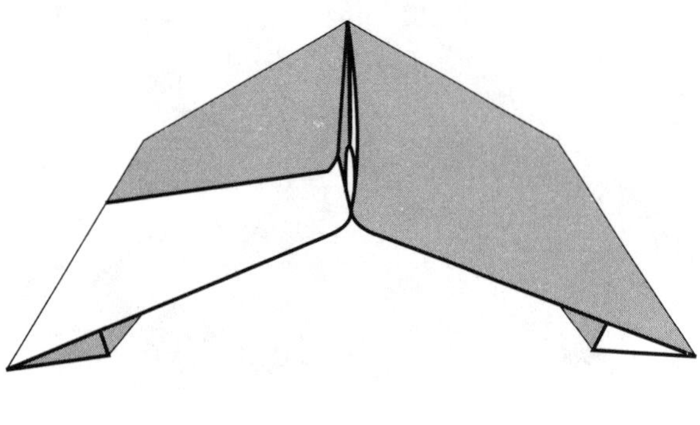

13. Now press the tip downward in the exact center, flattening the vertical triangular section against the rest of the plane, as shown in Fig. 14. Make certain to keep the central crease aligned, and crease down starting from the tip backward and from the center out.

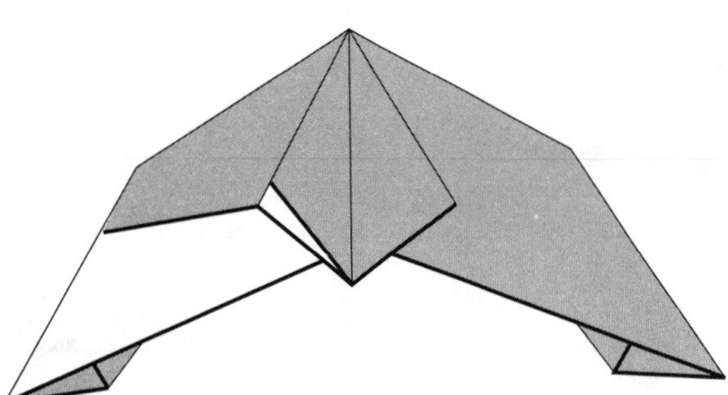

14. Shows the resulting plane from above. Fold in half along the central crease, matching the wings exactly against each other. Crease the wing edges together, flattening them, then open the plane out.

Flying instructions for PLANE 19 are on page 105.

Dart

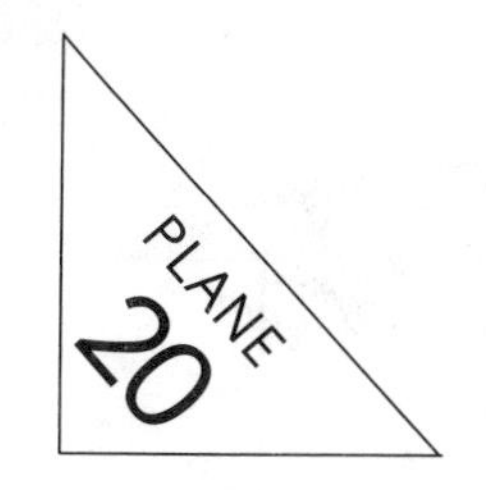

THE FINAL JET is also the most difficult—perhaps the most challenging paper airplane in this book, but what a plane. Almost as tricky to throw as to fold, the plane nonetheless flies very well, as befits the appearance. If you are just starting, try something else first. If you are up to the challenge, follow the directions very carefully and you might be delighted with the result. (Folding instructions are on pages 112–119.)

To fly

Note that the body of the plane is above the wings, rather than below. Hold the plane by the body from above with your palm toward the tip. Throw it forward with a brisk underhand or sidearm toss. Alternately, you can throw the plane with the body down (upside-down) and allow it to right itself in flight.

Adjustments

If you have folded the plane well, relatively little adjustment is necessary. Make certain the wings are symmetrical. You can steer the plane by adjusting the back edge of either set of wings—those in front (the canard) will act in reverse, however; if you bend them up, the plane will nose down, and vice versa.

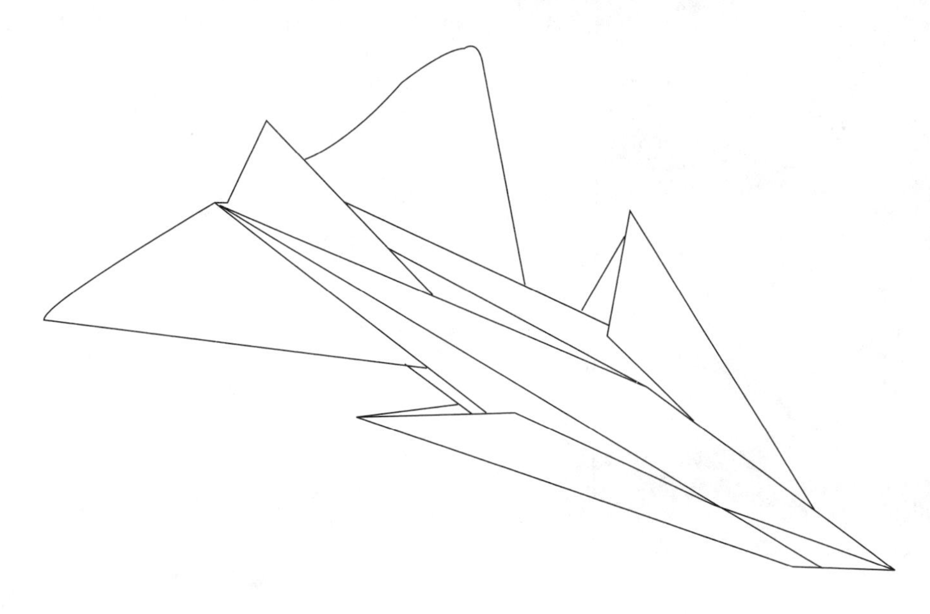

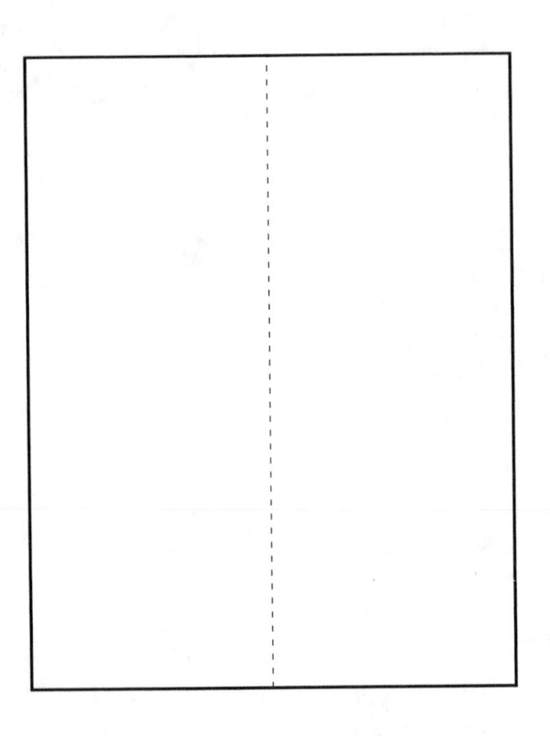

1. Make a central crease by aligning the left edge of the paper with the right. Crease well.

Unfold.

Turn the paper over.

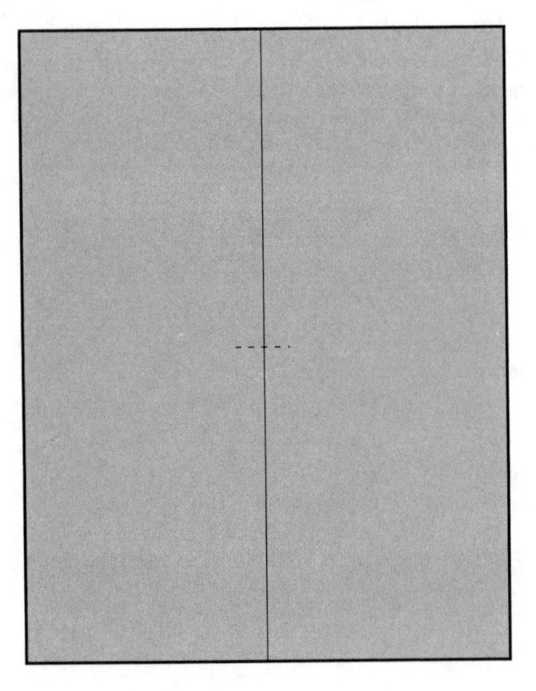

2. Mark the center of the central crease by aligning the top end of the crease with the bottom, and making a small crease in the center with your finger. Unfold.

3. Now fold down one end of the paper so it just meets the center mark you made in Step 2. Make certain the center crease is aligned, and crease very well.

Unfold.

Turn the paper over.

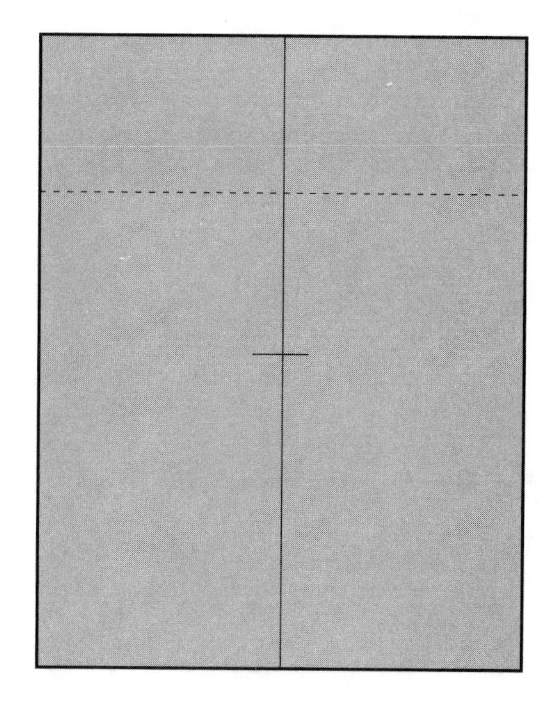

4. Now make crossing diagonal creases as shown. First fold down the left upper corner so that the fold passes through intersection C, and line A is aligned with the central crease.

Unfold.

Repeat with the right upper corner, aligning line B with the central crease. Again, your fold should pass through intersection C.

Unfold.

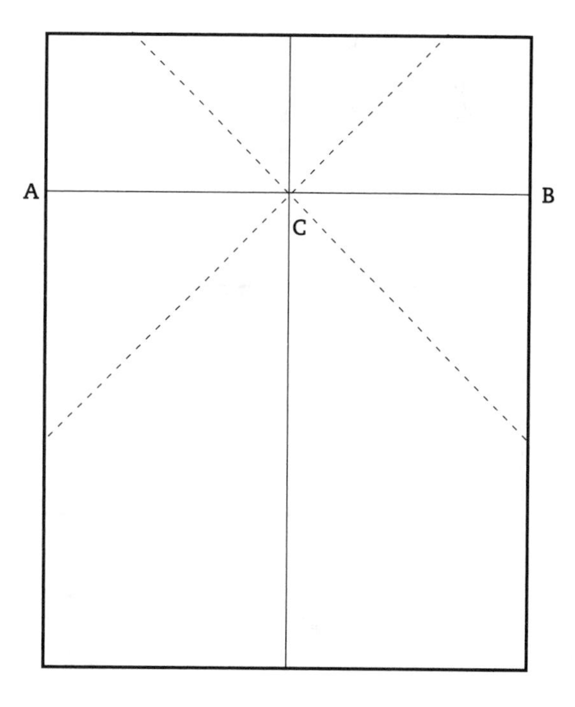

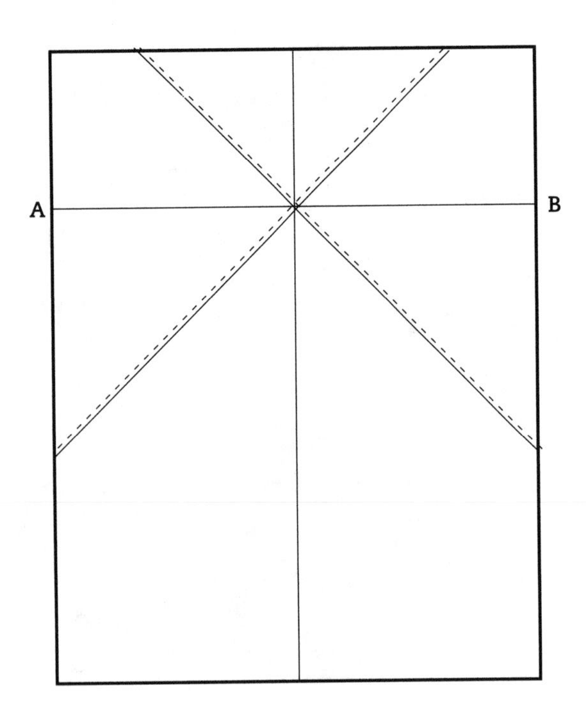

5. Now make an accordion fold by bringing point A and point B downward to lie against the central crease. The paper should fold along the diagonal creases, forming small upper wings to rest against the larger lower ones, as shown in Fig. 6. Recrease all folds, making certain that the upper wings align perfectly with the lower ones.

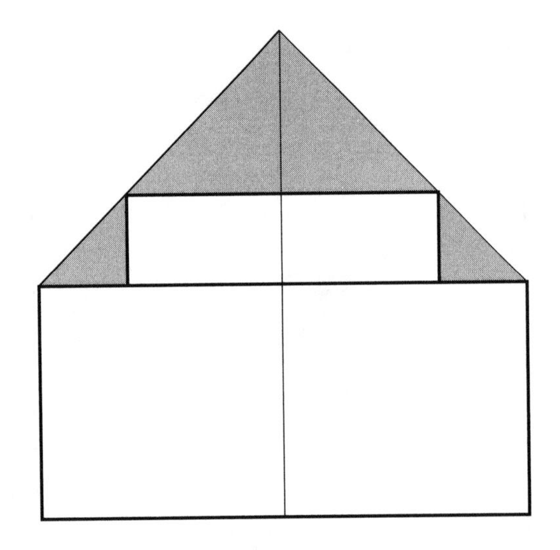

6. The result should look like this. Now fold the small upper wing on the right over to the left side.

7. Make a diagonal fold on the right, extending from the front tip of the plane to the right lower corner, as shown. Crease well.

Fold both small wings over to the right.

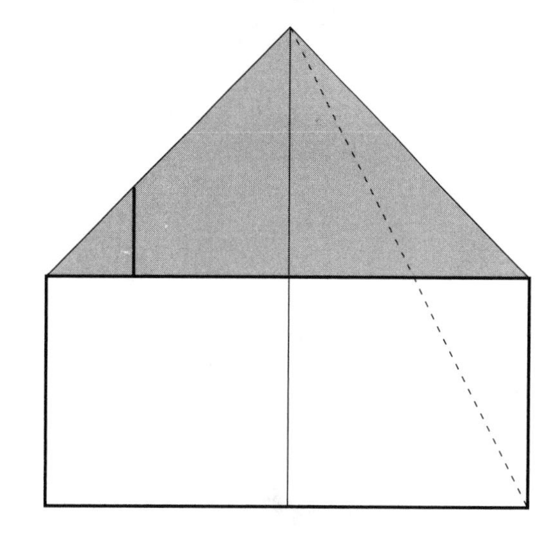

8. Make a diagonal fold on the left, again extending from the front tip to the left lower corner, as shown. Crease well.

Fold the plane in half along the central crease.

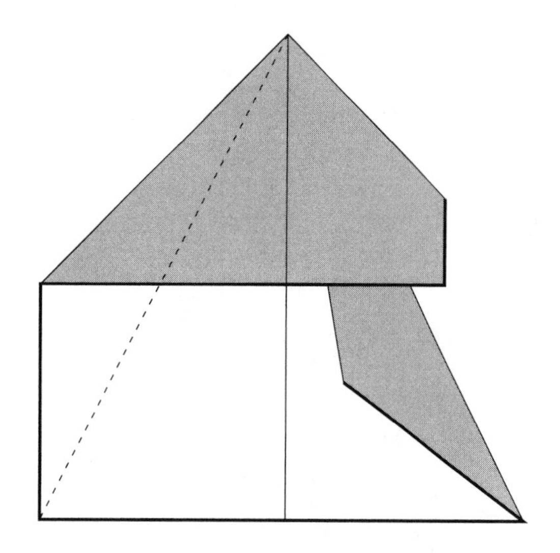

9. Fold down the large lower wings as shown, making your fold approximately 1½ inches from the central crease and parallel to the crease. (Align the back edge of the wing with the back edge of the body to make the fold perfectly parallel.)
(Step 9 continued on p.116)

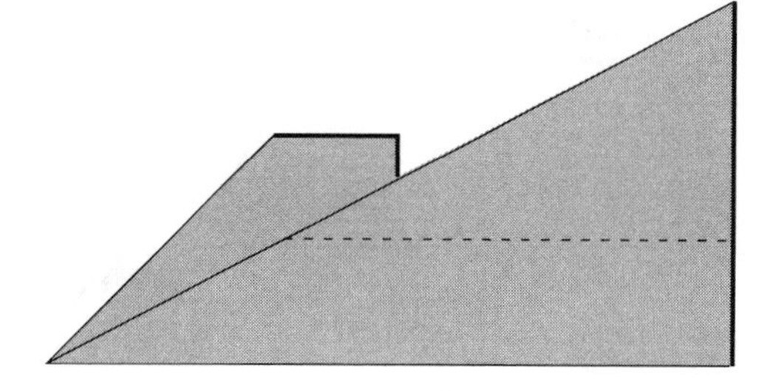

(step 9 continued from p.115.)

Make certain both wings are folded exactly the same, and crease well.

Open the plane out again, so that both small wings are folded over to the right, as shown in Fig. 10.

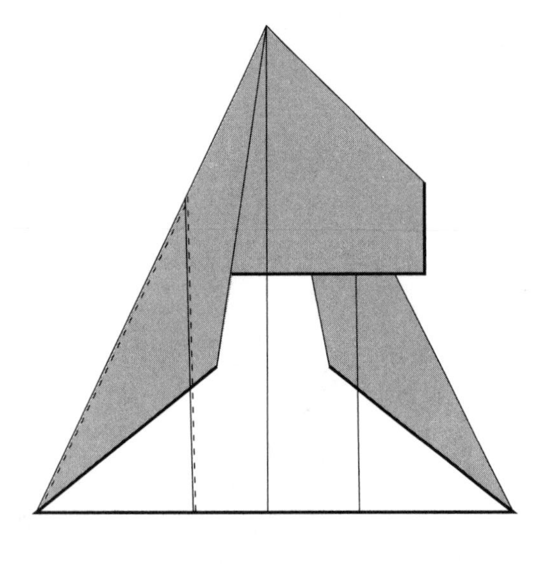

10. Now make an accordion fold as shown, by turning the outer tip of the left wing inside-out, folding it in along the vertical crease as shown. Crease well. The result should look like Fig. 11.

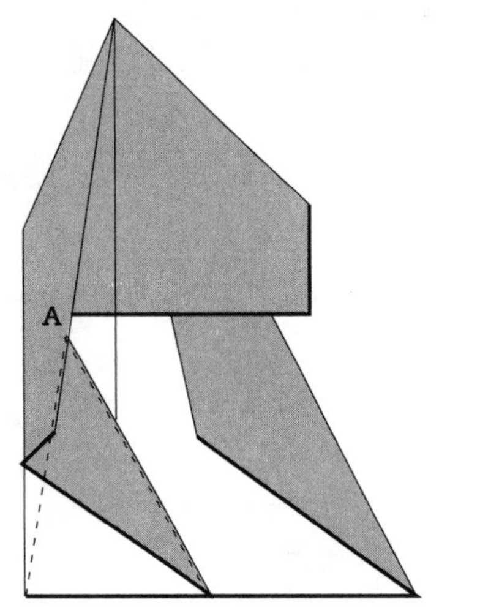

11. Make a second accordion fold on the same wing, turning the tip back, right side out. Start this fold at the point where the inverted wing crosses the horizontal crease, indicated by A.

Fold both small winglets over to the left. The result should look like Fig. 12.

12. Now make an accordion fold on the right, as you did on the left in Step 10. Crease well.

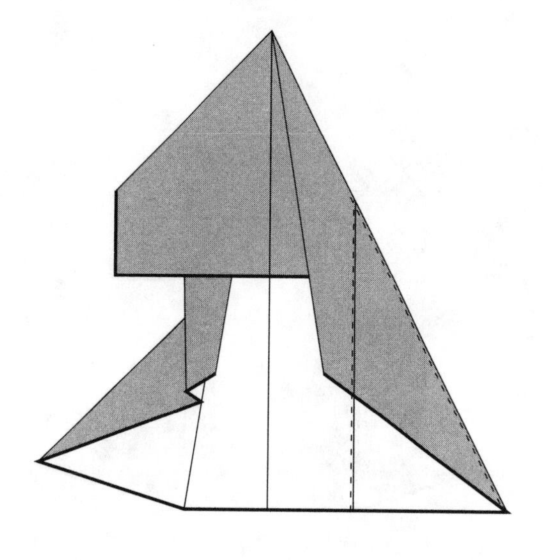

13. Make a second accordion fold on the right as you did on the left in Step 11, again folding from point A back to the tip. Crease well.

Fold the topmost winglet back over to the right, so that there is one on each side.

Turn the plane over.

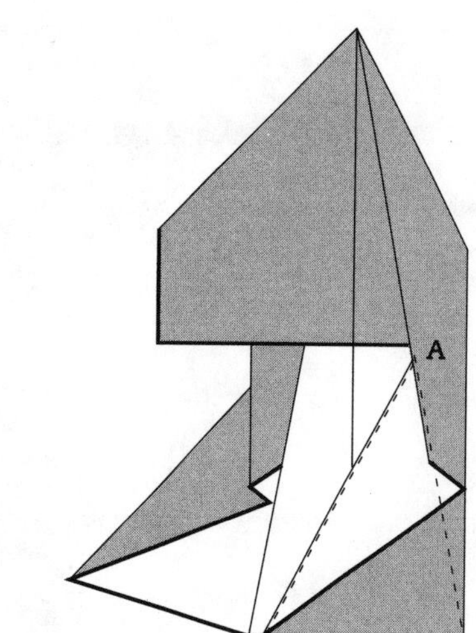

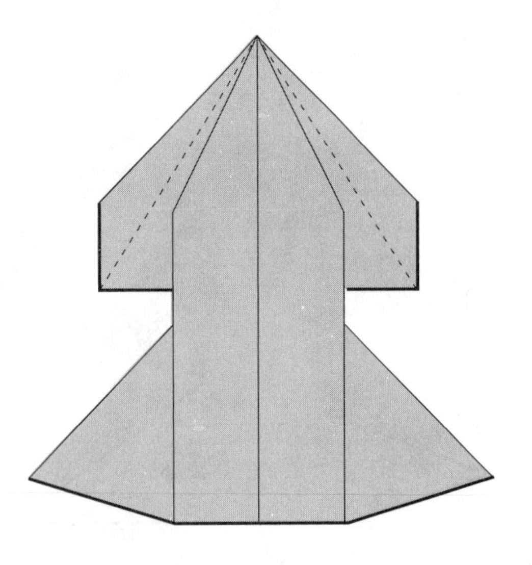

14. The plane should appear as shown. Make a diagonal fold on each winglet, extending from the tip of the plane to the lower corner of the winglet on each side, as in the drawing.

Turn the plane over.

Fold in half along the central crease. Make certain both front and back wings align perfectly on each side.

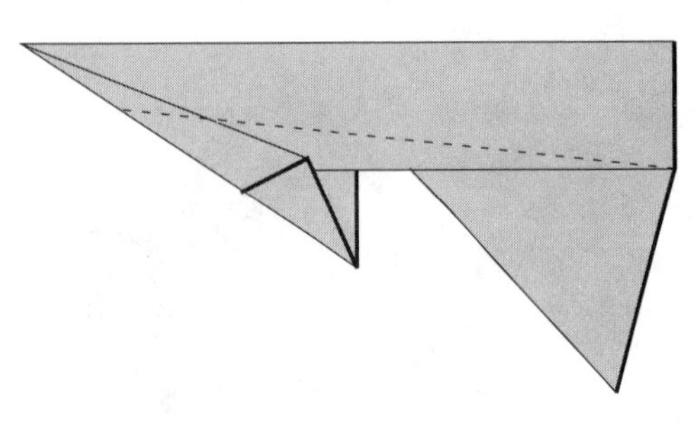

15. Now fold up both front and back wings on each side as shown. The fold should start from the base of the wings in the back, and taper toward the front until it is approximately ¾ inch from the central fold. Make certain the wings on both sides are exactly matched with each other, and crease well.

Unfold the wings.

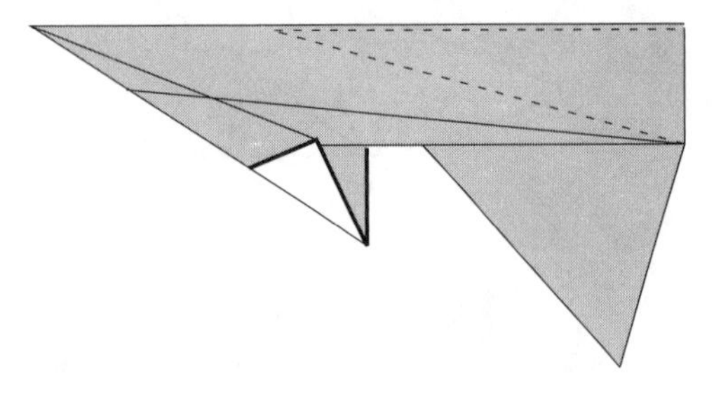

16. Fold the tail down in an accordion fold as shown, turning the tail section inside out, as in Fig. 17. (It might help to make this fold to one side first, unfold, then refold it as an accordion fold.)

17. This demonstrates the tail fold from below, showing how far forward the accordion fold should reach.

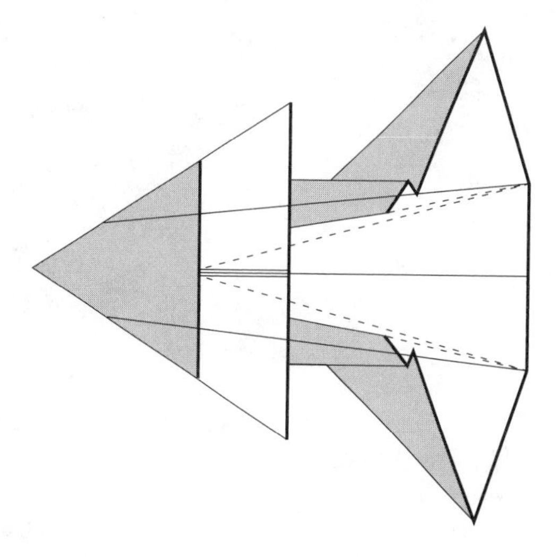

18. This shows a side view with the wings folded upward. Fold the tail back upward in a second accordion fold, with your fold following the bottom of the plane as shown. The tip of the tail should be brought upward between the wings, turning the entire section inside out. Crease well.

19. Shows the finished plane, from above. Recrease all folds, and make certain both wings are as exactly alike as possible.

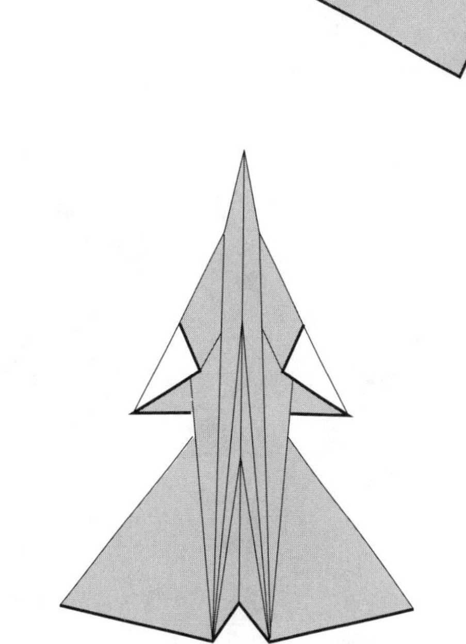

Flying instructions for PLANE 20 are on page 111.

Helicopters

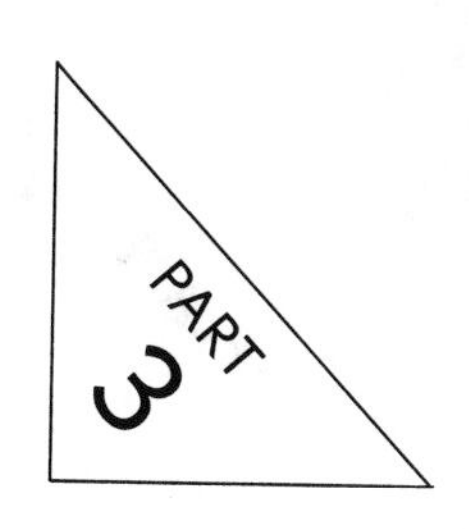

A DIFFERENT breed of paper airplane—designed not to soar but to spin in the air, much like a windmill or an unpowered helicopter—comprises this last part. The principle behind the paper helicopter is simple—rather than folding a plane with a right and a left wing, as most airplanes have, you fold a helicopter with two right wings, or with two left wings. Having been duped in this way, the plane is unable to fly in a straight line, but instead must chase its own tail endlessly through the air; hence, the spinning motion.

Basic helicopter

THIS OFFBEAT design is no more than the helicopter concept scaled down to its simplest form. Exceedingly easy to fold, it nonetheless spins quite well, and provides a good introduction to this unorthodox section. (Folding instructions are on pages 123–124.)

To fly

Hold above your head with the fold pointing down. Release, and stand aside, allowing the helicopter to spin to the floor.

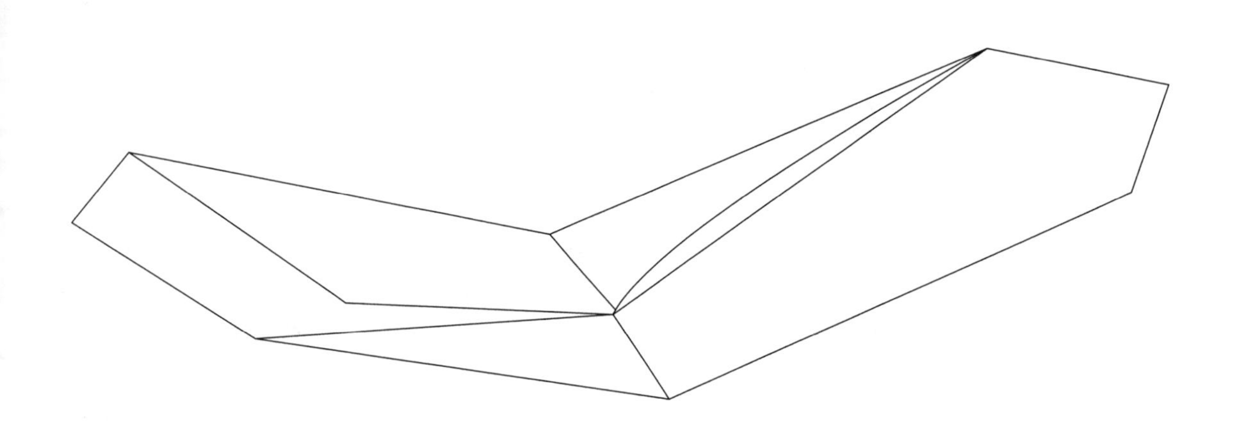

1. Make a diagonal crease as shown, by folding the right lower corner over to touch the left upper corner. Crease well.

Unfold.

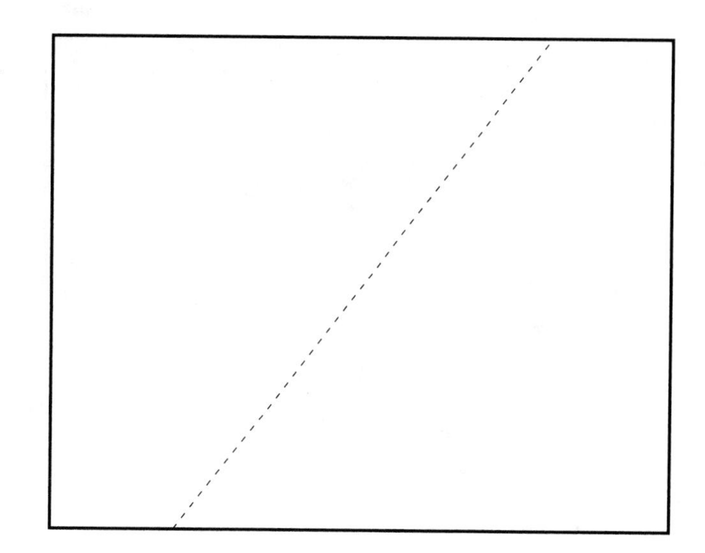

2. Make a fold starting from point A, aligning the top edge with the diagonal crease, as shown in Fig. 3.

Make a similar fold from point B, aligning the bottom edge with the diagonal crease.

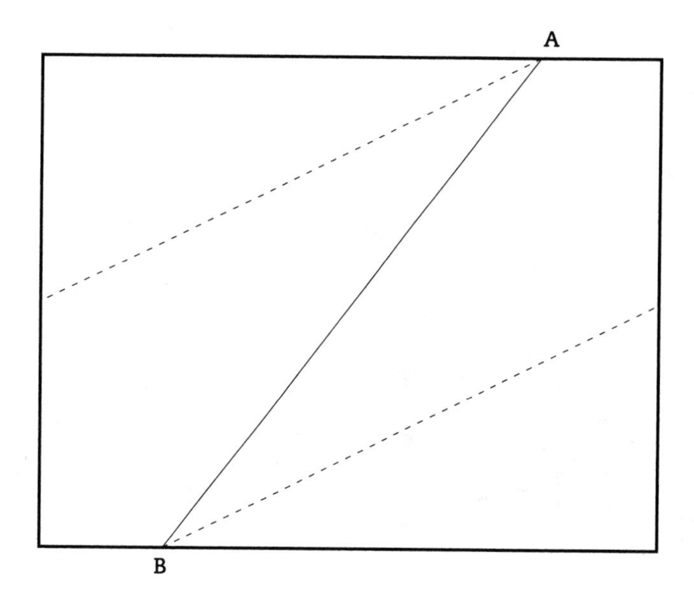

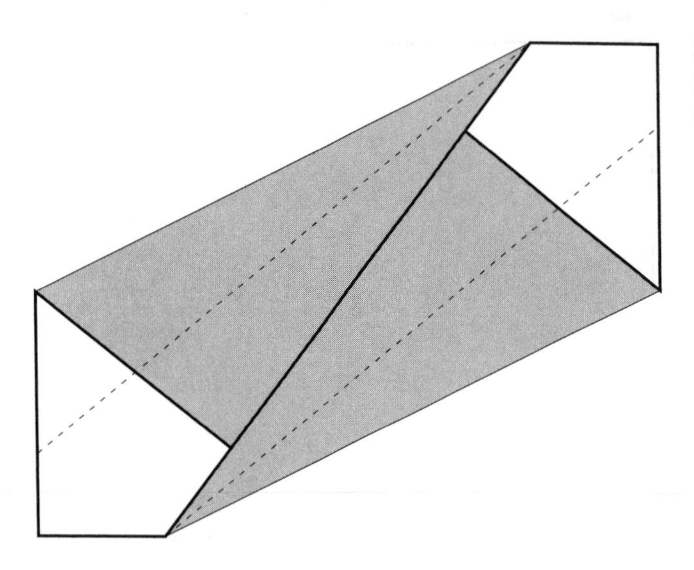

3. Shows the result of Step 2. Now make a second similar fold on each side of the diagonal crease, as shown.

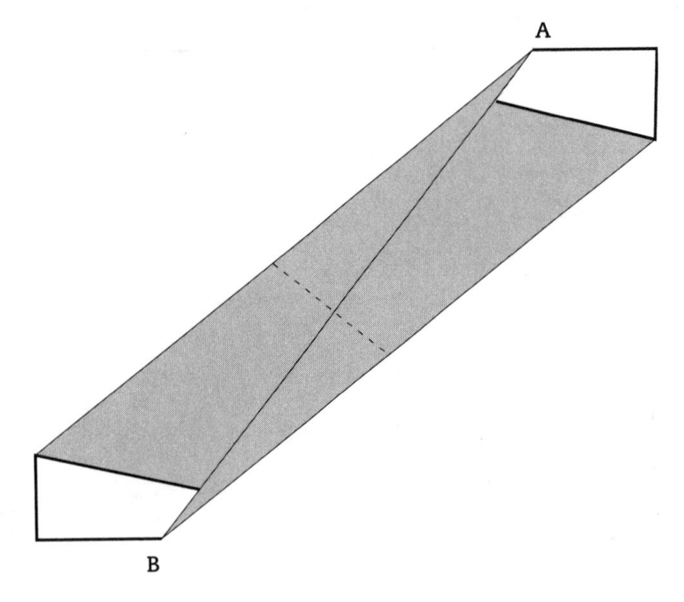

4. Now fold in half as shown, aligning point A with point B. Partially unfold. This is the finished plane.

Flying instructions for PLANE 21 are on page 122.

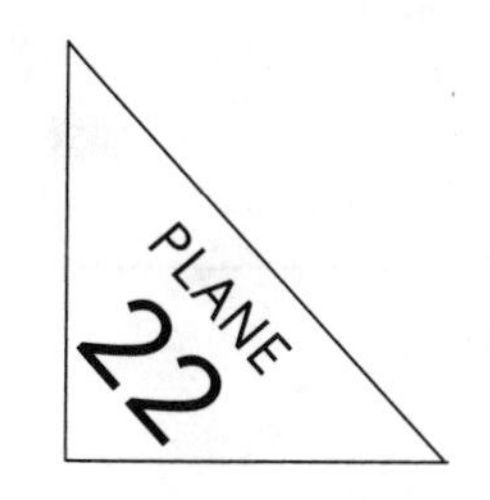

Broad-winged helicopter

THIS TWO-WINGED HELICOPTER is a slightly more involved design that spins much faster, yet is still simple to fold. (Folding instructions are on pages 126–130.)

To fly

Hold the aircraft from beneath and release point down.

Adjustments

The angles of the two wing folds in Step 10 are the key elements for adjusting this helicopter. The folds should be as even as possible on each wing. Experiment with the angles until you get a good spin.

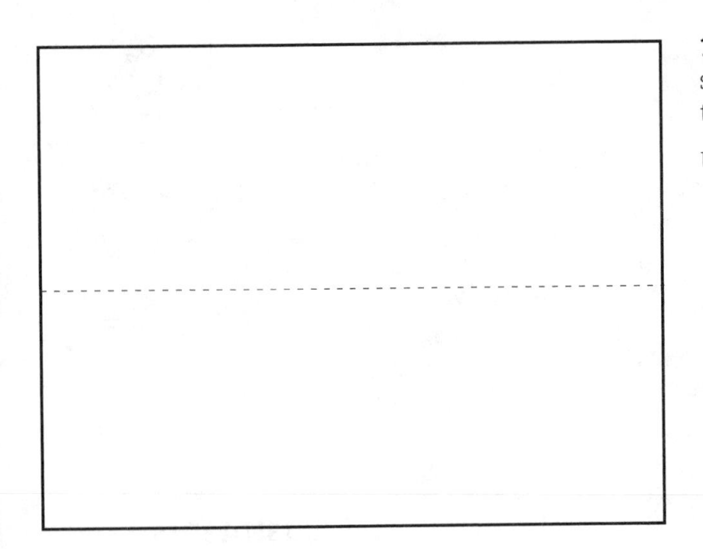

1. Make a horizontal central crease as shown, by aligning the bottom edge of the paper exactly with the top edge.

Unfold.

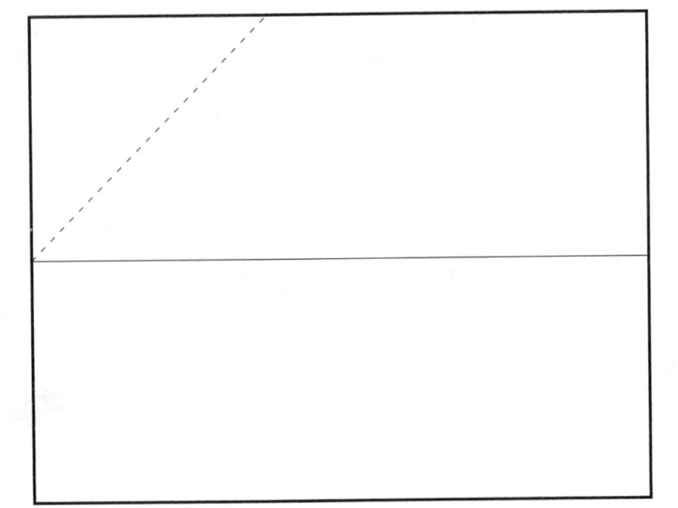

2. Fold down the left upper corner as shown, aligning the side edge with the central crease as shown in Fig. 3.

3. Make a second similar fold, again aligning with the central crease.

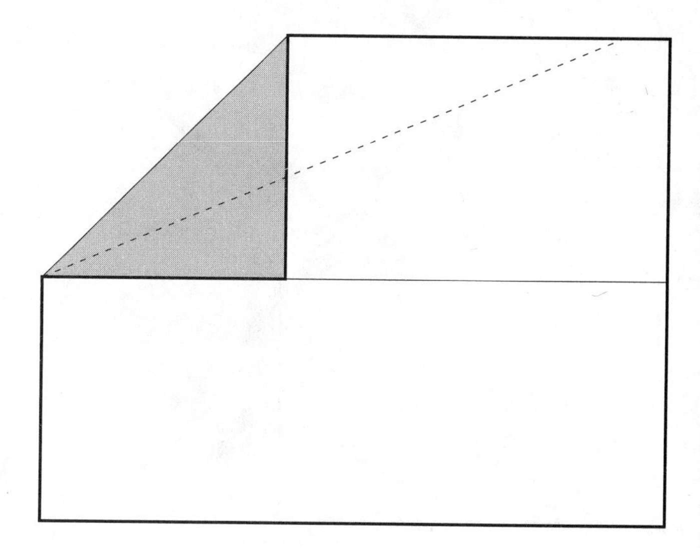

4. Now fold up the right lower corner in the same way, aligning the side edge with the central crease as shown in Fig. 5.

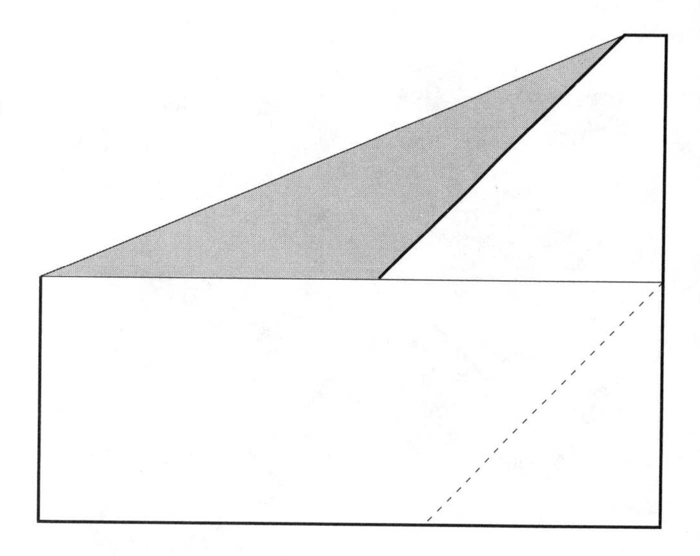

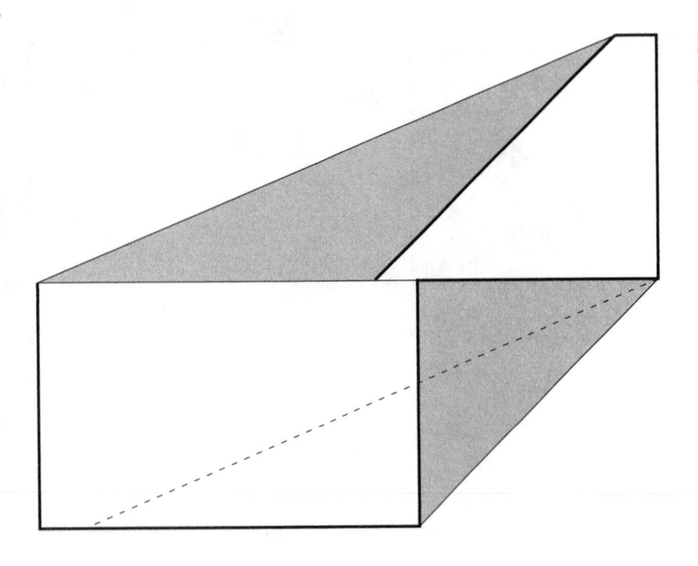

5. Make another similar fold, again aligning with the central crease.

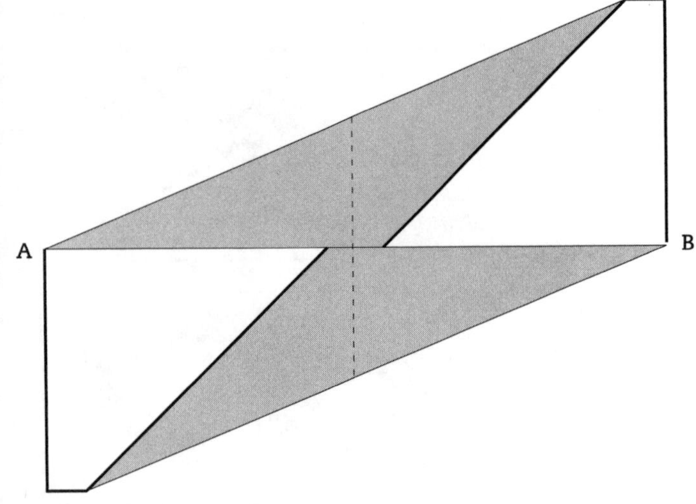

6. Fold the helicopter in half, aligning point A with point B.

Unfold. (Note: At this point, the aircraft will spin as a helicopter, and may be dropped according to the instructions for PLANE 21.)

7. Fold in half along the long central crease.

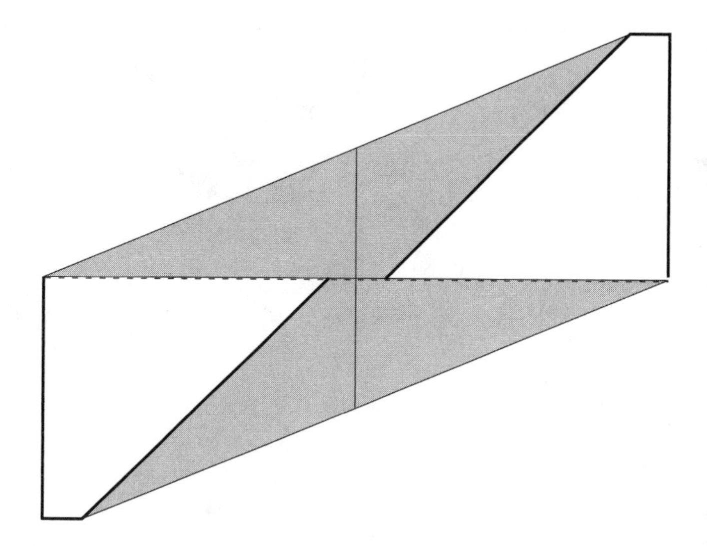

8. Fold the right wing upward as shown, aligning the bottom edge with the vertical central crease.

Turn the aircraft over, and position it as in Fig. 9.

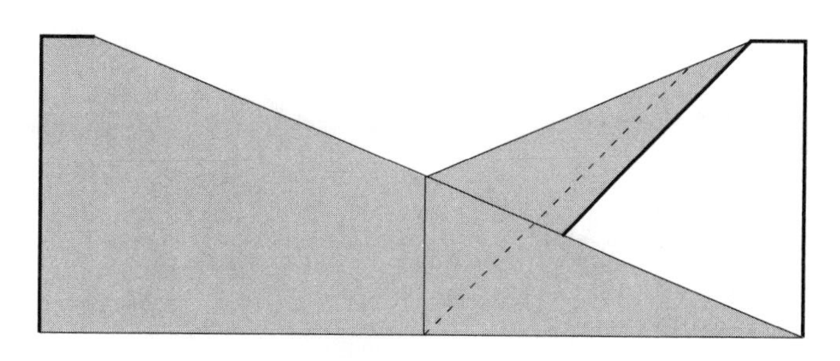

9. Fold up the other wing in exactly the same way, again aligning the bottom edge with the central crease. If you do this carefully, the aircraft should look like Fig. 10.

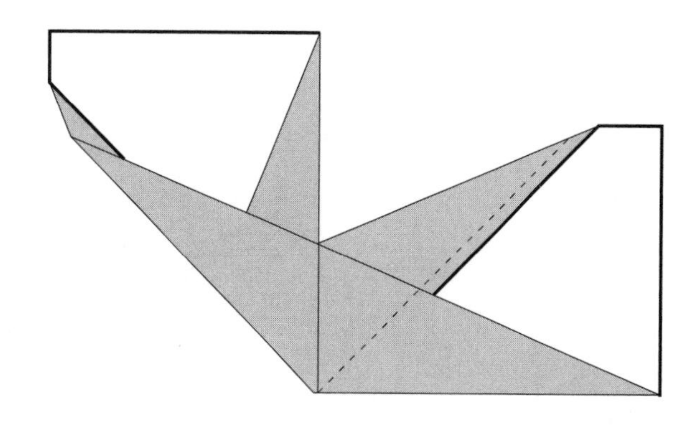

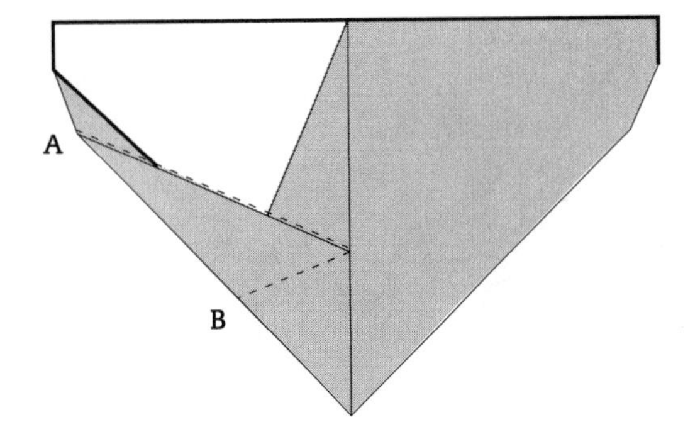

10. Fold down the wing on the left, folding over the edge that ends at point A.

Fold down also at point B. The wing will bend naturally along the line indicated.

Turn the aircraft over. Note that the other wing is now on the left.

Repeat the same two folds on that wing. Adjust the folds so that each wing is folded at approximately the same angle.

Flying instructions for PLANE 22 are on page 125.

Rabbit ears

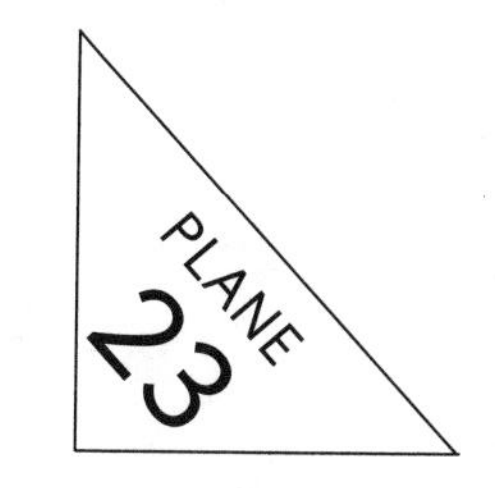

THIS HELICOPTER is named for long and narrow wings that ultimately produce a very gratifying spin. (Folding instructions are on pages 132–135.)

To fly

Hold from beneath by the point, and let the aircraft drop, stepping out of the way.

Adjustments

Adjust the angle of the wing folds until you get a satisfactory spin. Crease the pointed end very tightly to prevent the wings from unbending.

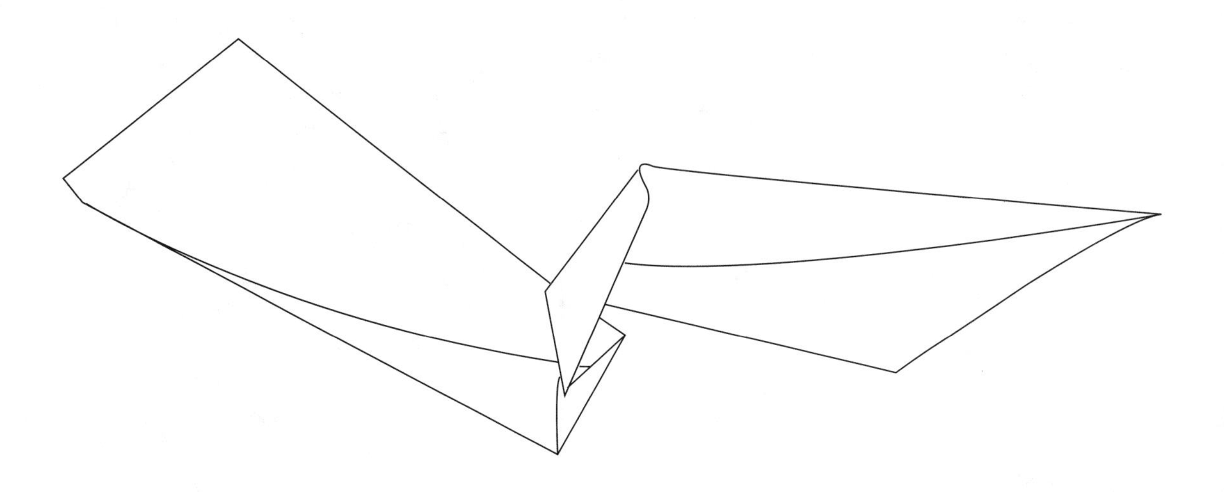

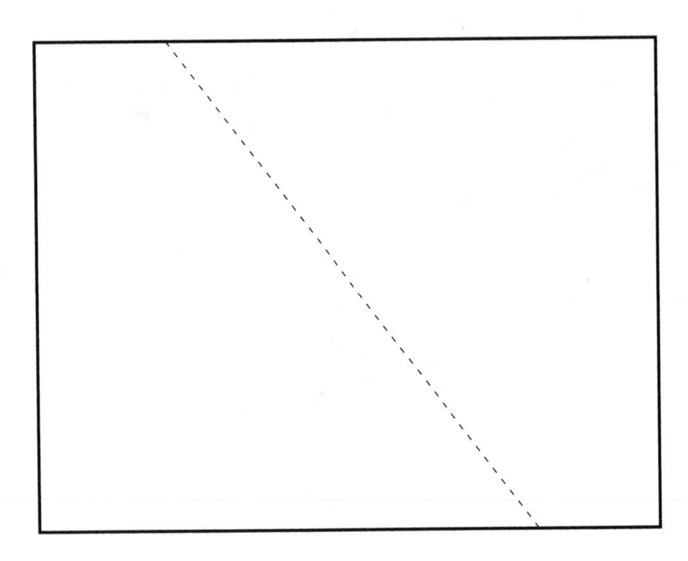

1. Make a diagonal fold as shown, by aligning the left lower corner of the paper with the right upper corner.

Unfold.

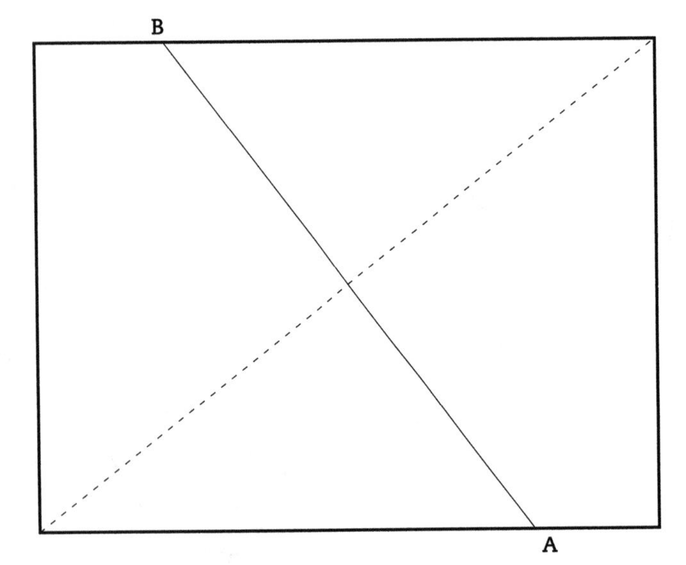

2. Make another crease crosswise to the first by aligning one end of the diagonal crease, marked A, with the other end, marked B, and creasing between them.

Unfold. If done correctly, your crease should extend from corner to corner.

3. Make a fold from the bottom left corner as shown, by aligning the bottom edge with the diagonal crease as shown in Fig. 4.

Make a similar fold from the top right corner, aligning the top edge of the paper with the diagonal crease.

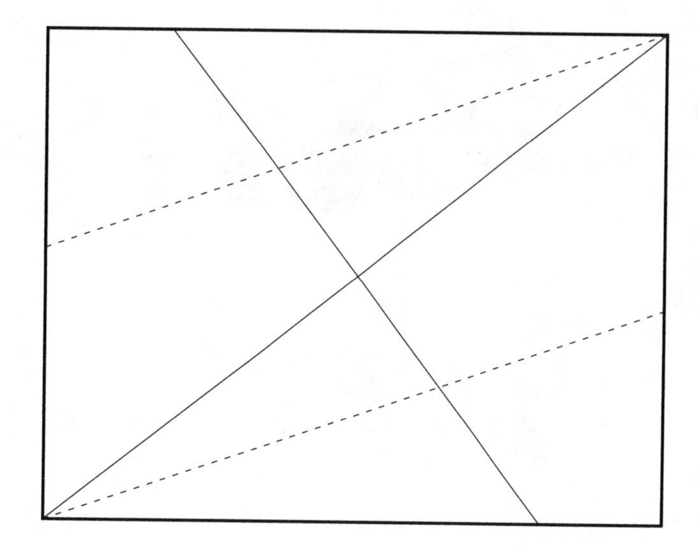

4. Make another similar crease from each side, as shown. The result should look like Fig. 5.

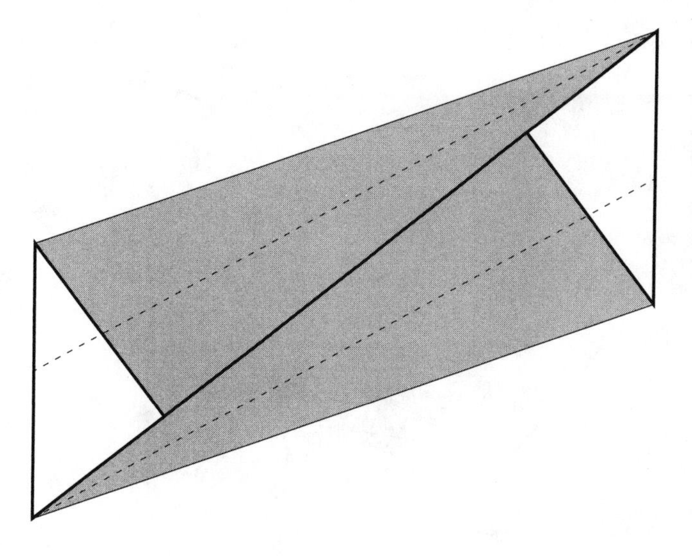

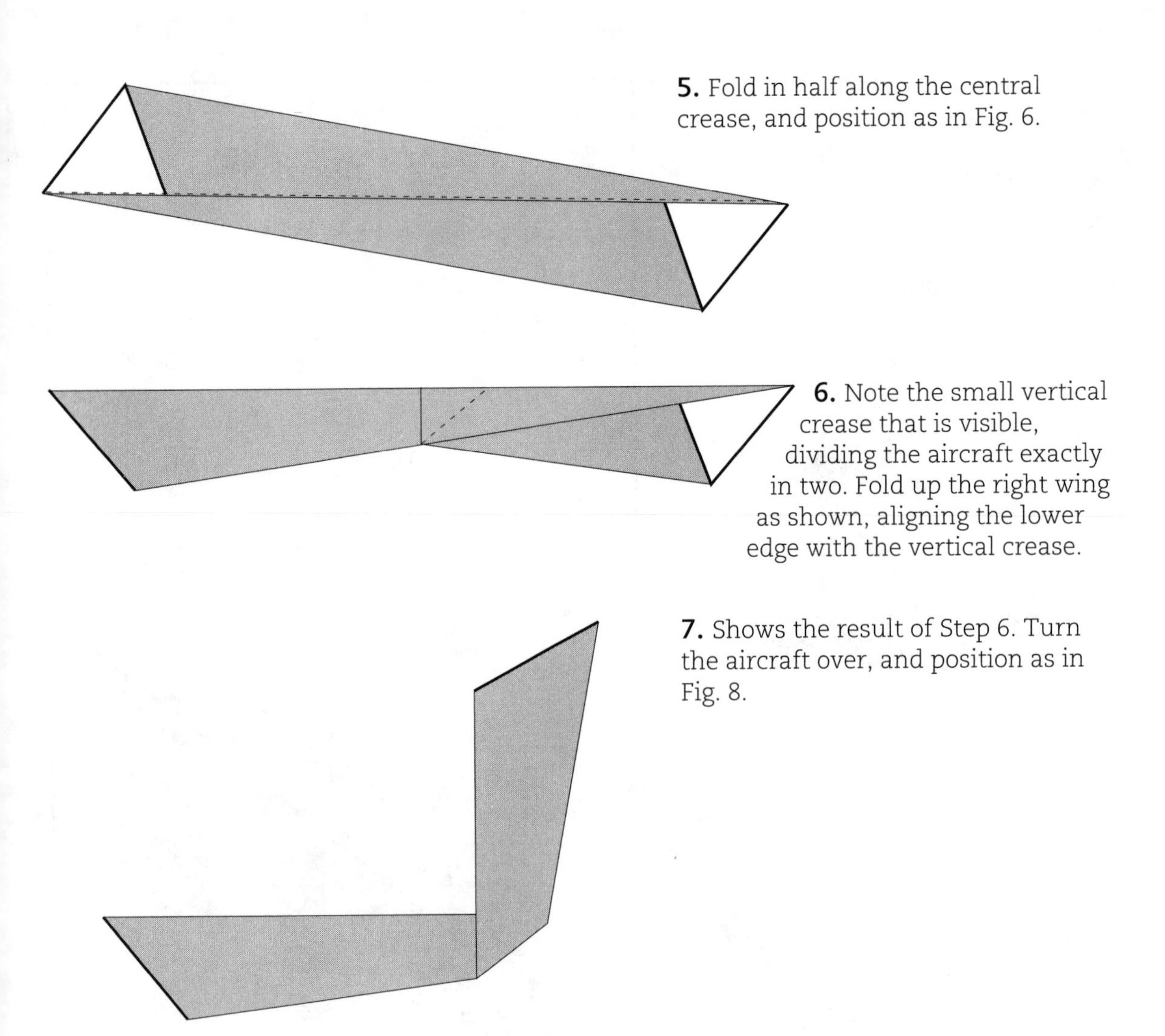

5. Fold in half along the central crease, and position as in Fig. 6.

6. Note the small vertical crease that is visible, dividing the aircraft exactly in two. Fold up the right wing as shown, aligning the lower edge with the vertical crease.

7. Shows the result of Step 6. Turn the aircraft over, and position as in Fig. 8.

8. Now fold the other wing up, again aligning the bottom edge with the small vertical crease. The result should look like Fig. 9.

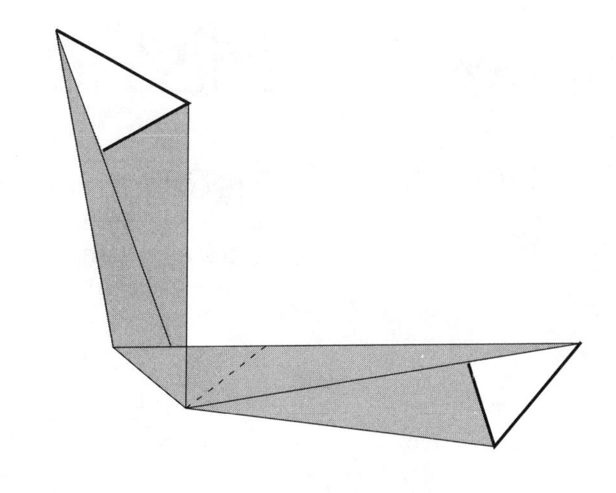

9. Fold down the left wing along the horizontal edge, as shown.

Turn the aircraft over. Note that the other wing is now on the left.

Fold this wing down as you did the first.

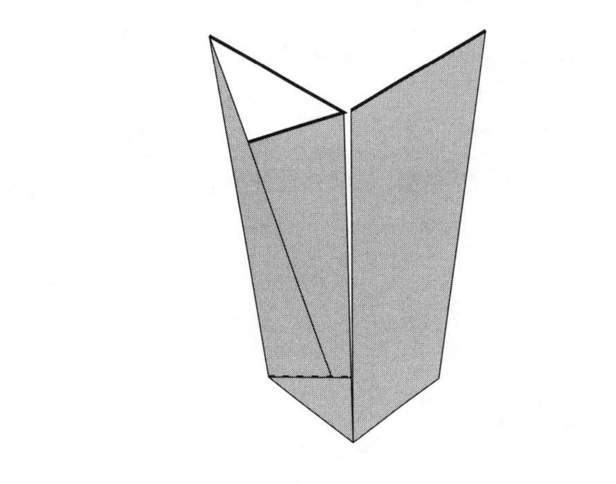

10. This shows how the finished aircraft should look, from above. Crease the edges of both wings to flatten them.

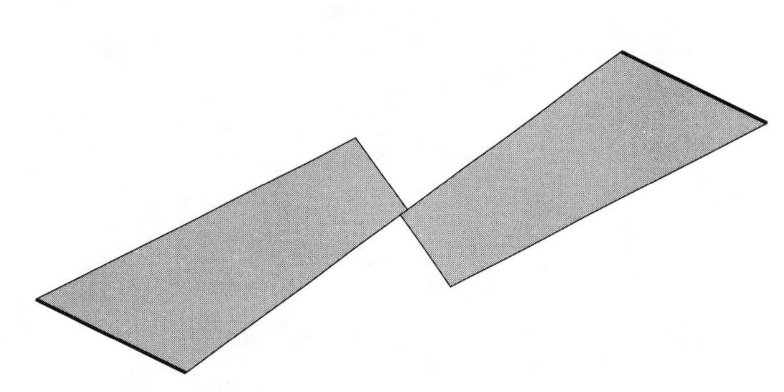

Flying instructions for PLANE 23 are on page 131.

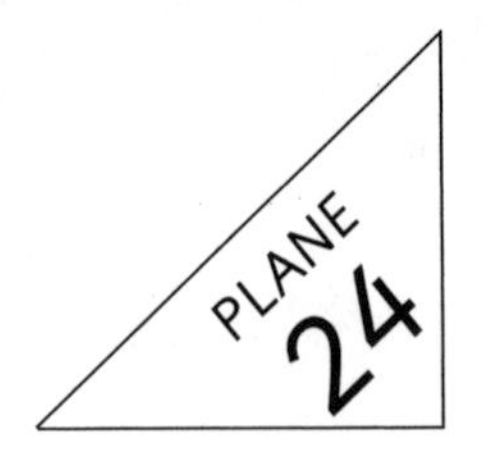

Missile

THROW THIS design through the air and it will fly much like an arrow, spinning as it goes. It is quite simple to fold and is a refreshing change from the falling helicopter designs. (Folding instructions are on pages 137–141.)

To fly

Hold the aircraft from beneath, toward the tip, and throw firmly tip-first, as though throwing an airplane. The aircraft will hold a relatively straight course, spinning as it goes.

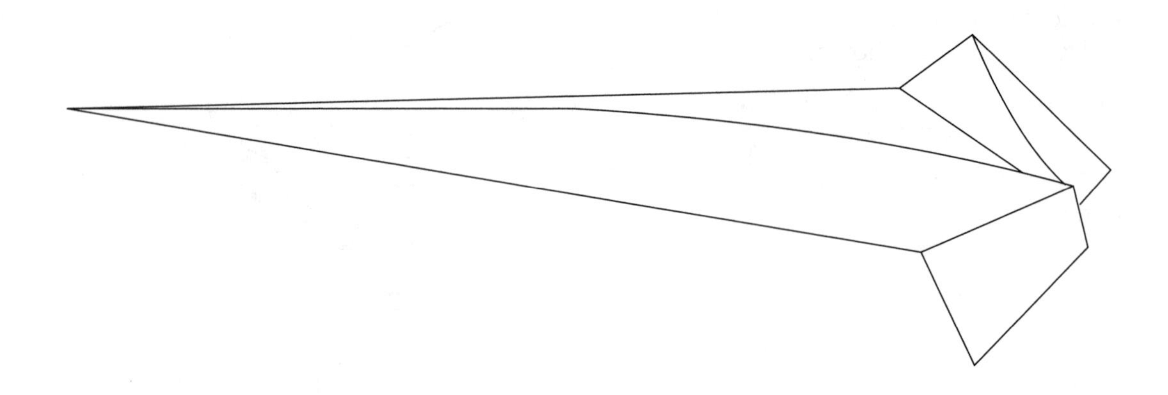

1. Make a vertical central crease as shown by aligning the left edge exactly with the right, and creasing between.

Unfold, and flatten the paper well.

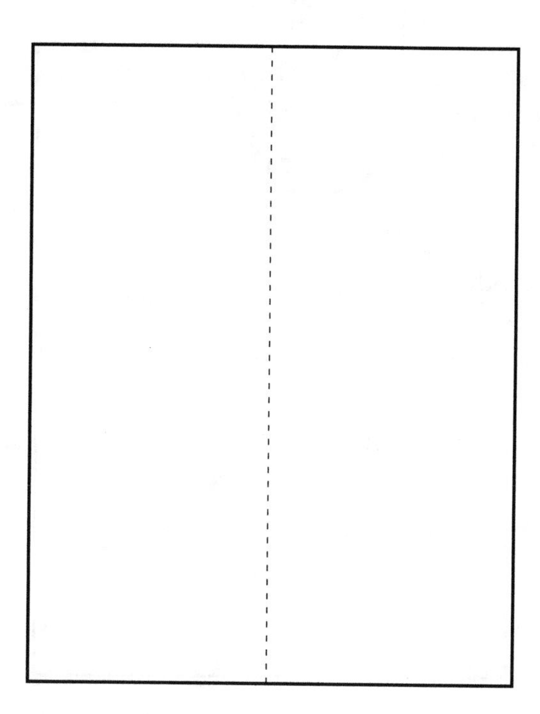

2. Fold down the left upper corner as shown, aligning the top edge against the central crease as shown in Fig. 3.

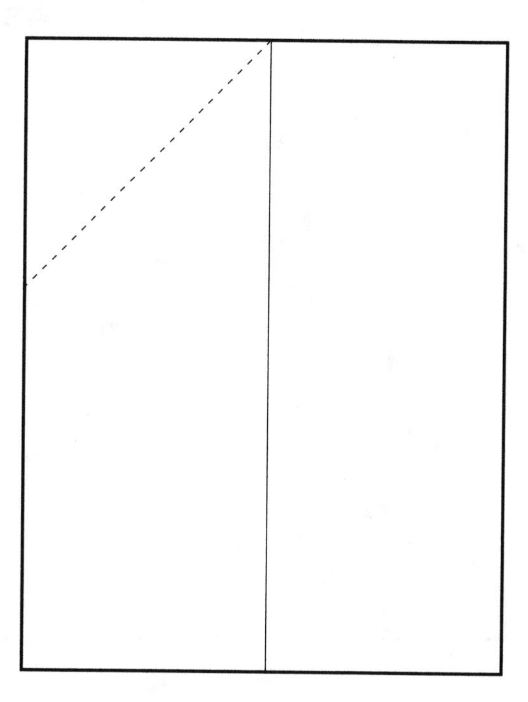

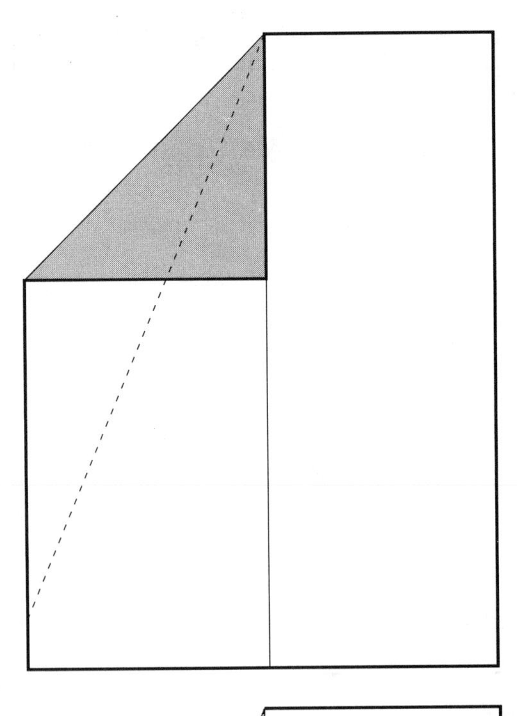

3. Make another similar fold, aligning the diagonal edge with the central crease, as shown in Fig. 4.

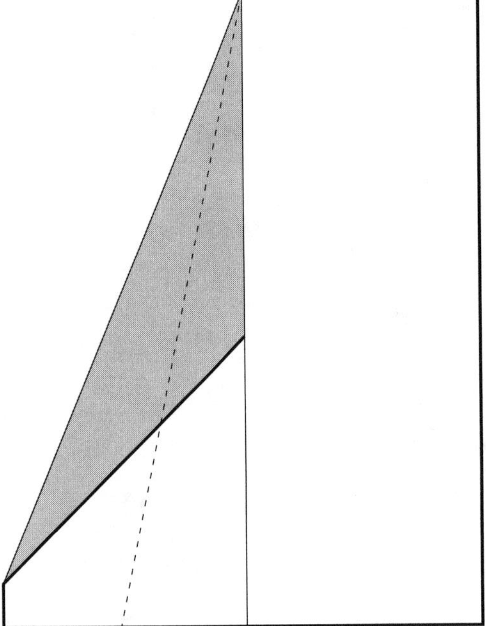

4. Make yet another fold, again aligning the diagonal edge against the central crease.

5. Shows the result of Step 4. Turn the paper over.

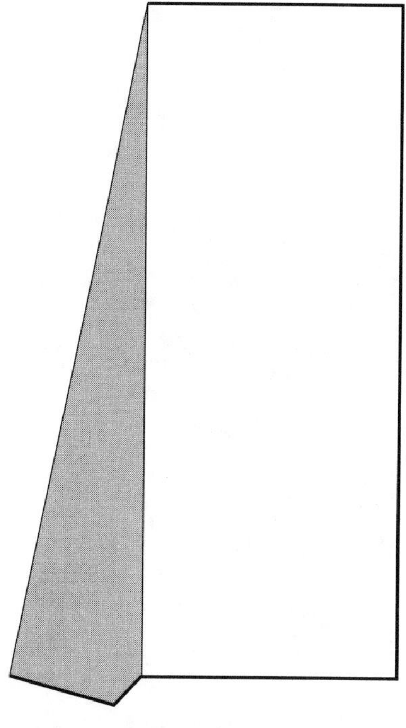

6. Now fold down the left upper corner again, as you did in Step 2.

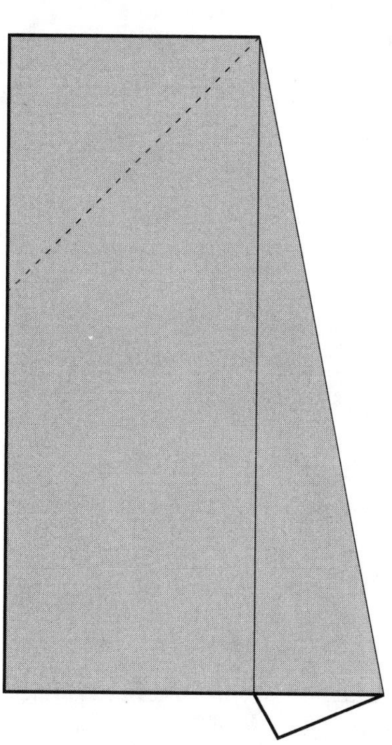

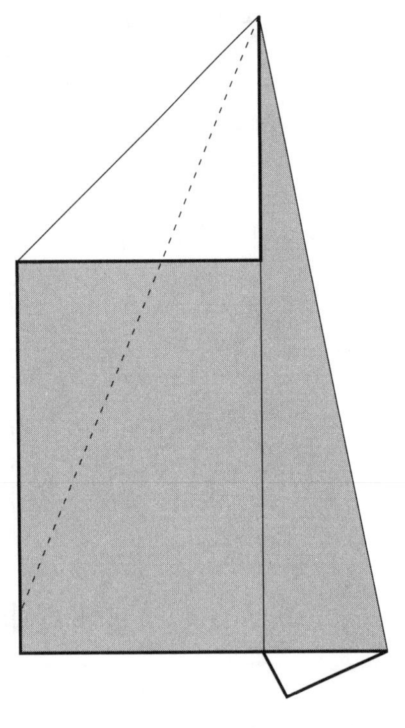

7. Fold down again, as in Step 3.

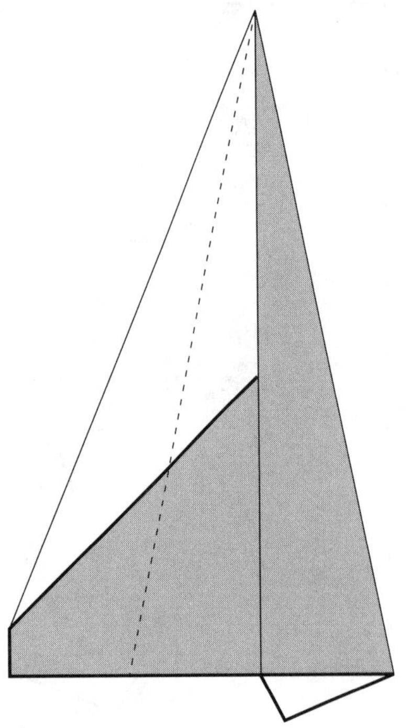

8. Fold down yet again, as in Step 4. The result should look like Fig. 9. Run the entire aircraft through your fingers, creasing all the edges as flat as you can. Then position the aircraft as in Fig. 9.

9. Fold down the tail on the left as shown, placing the corner marked A against the central crease.

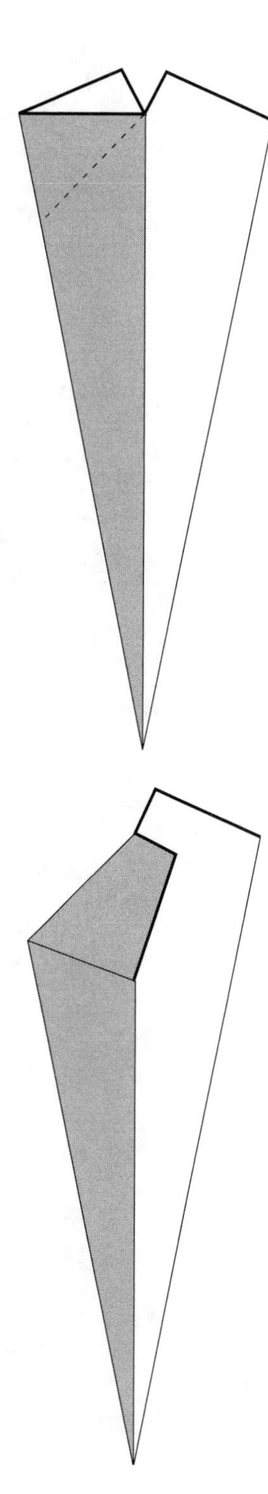

10. Shows the result of Step 9. Turn the aircraft over. Note that the other wing is now on the left.

Again fold over the tail in exactly the same fashion. Unfold the tail partially.

Flying instructions for PLANE 24 are on page 136.

INDEX

A
accordion fold, xiv-xv, 52, 79, 101, 116-119
aerobat, 31-35
aerobatic jet, 63-67
aileron steering, xvii
airfoil, 2, 6
airplane, essential, 2-5

B
batwing, 105-110
biplane, 50-56
broad-winged helicopter, 125-130

C
canard, 92
canard jet, 92-97
correct a turn, 81

D
dart, 111-120

E
essential airplane, 2-5
elevator steering, xvi

F
finned wings, 25-30
flying wing, 98-110
  batwing, 105-110
  stealth, 98-104
flying, xv-xvii
  proper throw, xv-xvi
folding,
  accordion, xiv-xv, 52, 79, 101, 116-119
  basics, xi-xv
  edge, against, xii-xvi
  paper, xi-xv
  point to point, xiv-xv
  rules, x-xi
  telescoping fold, xv, 53
fuselage plane, 40-49

G
glider, long-winged, 21-24

H
helicopter, 122-124
  broad-winged, 125-130
hornet, 58-62

L
long-winged glider, 21-24

M
midget, 35-40
missile, 135-141

N
Navy jet, 87-91

R
rabbit ears, 130-135
round wing, 16-20
rules, x-xi

S
shaping, proper, xvi
sleek jet #1, 73-80
sleek jet #2, 81-86
stealth flying wing, 98-104
steering, proper, xvi-xvii
  aileron, xvii
  dives, xvii
  descends, xvii
  elevator, xvii
  noses up, xvii
  veers off, xvii
straight wing, 6-10

T
telescoping crease, 54
telescoping fold, xv, 53
turn, correction, 81
twin fin, 68-73

U
underside flaps, 11-15